Numerical Solution
of
Differential Equations

This is a volume in

COMPUTER SCIENCE AND APPLIED MATHEMATICS

A Series of Monographs and Textbooks

Editor: WERNER RHEINBOLDT

A complete list of titles in this series appears at the end of this volume.

Numerical Solution
of
Differential Equations

Isaac Fried

Department of Mathematics
Boston University
Boston, Massachusetts

QA
371
.F795

ACADEMIC PRESS   New York San Francisco London   1979
A Subsidiary of Harcourt Brace Jovanovich, Publishers

ACADEMIC PRESS, INC.
111 Fifth Avenue, New York, New York 10003

United Kingdom Edition published by
ACADEMIC PRESS, INC. (LONDON) LTD.
24/28 Oval Road, London NW1   7DX

Library of Congress Cataloging in Publication Data

Fried, Isaac.
   Numerical solution of differential equations.

   (Computer science and applied mathematics)
   Bibliography: p.
   1. Differential equations––Numerical–solutions––
Data processing. 2. Difference equations––Numerical
solutions––Data processing. 3. Finite element method––
Data Processing. I. Title.
QA371.F795   519.4   78–26454
ISBN 0–12–267780–3

PRINTED IN THE UNITED STATES OF AMERICA

79 80 81 82 83 84 9 8 7 6 5 4 3 2 1

מֵאָדָם עַד נֹחַ

Contents

4 *FINITE ELEMENTS*

5 *DISCRETIZATION ACCURACY*

6 *EIGENPROBLEMS*

10 WAVE PROPAGATION

Preface

Occupying a unique place along the border between applied mathematics and the concrete world of industry, the numerical solution of differential equations, probably more than any other branch of numerical analysis, is in a constant state of unrest and evolution. Being so widely and variously applied in the real world, its techniques are relentlessly put to the ruthless test of practical success and usefulness. Nor does it evolve solely through the cross influences of the practical necessities of engineering; unusual impetus is also given to this field by the astounding advances in computer technology, which is gathering now to miniaturize hardware to lower the cost of the equipment, the arithmetic, the logic, the storage, and the output that is made more comprehensively grasped by directly presenting it to that most remarkable of the human senses—vision, through computer graphics, shifting thereby the engineer's or programmer's priorities in selecting the most appropriate solution algorithm.

Discretization of a differential equation, or for that matter the physical process behind it, in preparation for the programming of its solution is greatly responsible for the surge in interest in finite mathematics and particularly linear algebra as the mathematical link between the continuous formulation of the problem and the properties of the computer chosen to solve it. Linear algebra is now making its way from the periphery to the center of the undergraduate curriculum in the sciences, and my underlying assumption in writing this book was that the reader is familiar with computational linear algebra.

Several outstanding textbooks, some highly theoretical, others more technically inclined, are available on the numerical solution of differential equations; but they either concentrate on the initial value problem, which requires only a modicum of linear algebra, or they almost exclusively deal, as is the case mostly, in partial differential equations with boundary value problems in two or three dimensions. What, it is hoped, makes this book different is that it includes at once the initial, boundary, and eigenvalue problems. The book is also more emphatic on the relationship among the analytic formulation of the physical event, the discretization techniques applied to it, the algebraic properties of the discrete systems created, and the properties of the digital computer. It is the proper interaction between these components of the solution plan that decides its ultimate success. I have put strong, maybe inordinate for the circumstances, emphasis on the most-recent exciting and influential invention in computational mechanics–finite elements. This was done partly for its intrinsic usefulness in the numerical solution of differential equations and its theoretical soundness, and partly in order to fully introduce this remarkable method in a setting simpler than partial differential equations where its power is much more apparent but where the technique is decidedly the same as for ordinary differential equations.

The book is intended as an introduction to the numerical solution of differential equations but proceeds to embrace all that I deemed useful for the person interested in the actual application of the techniques. Starting from the fundamentals, it goes on to present the principal useful discretization techniques and their theoretical aspects but without undue lengthy and pedantic mathematical arguments, and it includes ample geometrical and physical examples, mainly from continuum mechanics, to enhance the reader's practical perspectives. Huge effort has been put into the development of the techniques of the numerical solution of differential equations and their analysis. But even so, when it comes to applying them in a realistic engineering situation, the most theoretical analysis can offer is broad guidelines, insights, and general advice, rarely a concrete decisive answer. Much of the success of the numerical analyst rests with the combination of theoretical knowledge and skill that comes from experience.

The ultimate aim, of course, of all the techniques for the numerical solution of differential equations is to render the described physical problem programmable on a digital computer. This can be done either directly from the discretization of the basic physical principles, skipping the stage of analytical modeling and with disregard to the differential equation; and there are engineers who prefer it this way, claiming that their responsibility is to physical reality rather than to some simplified differential equation. Or the starting point of the discretization can be the differential equation with the appropriate boundary conditions in which differentials are replaced by finite differences. This latter approach is more widespread and accepted, maybe for historical reasons mainly; but there has been accumulated on differential equations a great wealth of information which can be put to great use in the analysis of their discretization. Also, the finite difference or finite element approximation permits greater sophistication in the discretization than shear physical discrete modeling; I followed this formal discretization procedure of differential equations throughout the book.

A complete list of references on the topics covered in this book could fill by itself several hefty volumes, and I restricted it therefore to a mere list of books only. As for the notation, no special typographical distinction is made between scalars, vectors, and matrices; the reader is judged keen enough to distinguish among them contextually.

Thanks are due to my graduate student Arthur R. Johnson who proofread the manuscript.

Numerical Solution
of
Differential Equations

1 *Finite Differences*

1. Calculus to Algebra to Arithmetic

At the heart of all methods for the computer solution of differential equations that we have in mind to discuss in this book lies the idea of computing with small finite, rather than infinitesimal, quantities and the postponement of the thrust to the limit, so typical of calculus, to the numerical stage of the solution, instead of carrying it out at the stage of the analytical formulation. Before proceeding to apply this idea to differential equations, we find it instructive to scrutinize its working in a more familiar setting.

Consider, for that purpose, the problem of computing the area A between the curve $y = \cos x$ and $y = 0$ for $0 \leqslant x \leqslant \pi/2$. First we wish to apply calculus to the solution of this problem and effortlessly write

$$A = \int_0^{\pi/2} \cos x \, dx = \sin(\pi/2) - \sin(0) = 1 \tag{1.1}$$

The solution seems simple to us, but the simpler Eq. (1.1) is the more astounded need we be at the power and conciseness of the symbol manipulation of calculus. Once we remember that $\sin x$ is the antiderivative of $\cos x$ the computation of A is reduced, according to the fundamental theorem of integral calculus, to the mere evaluation of $\sin x$ at $x = 0$ and $x = \pi/2$.

It now occurs to us to engage the digital computer to compute the area under $\cos x$, or for that matter, any other given curve. But by its very nature, the digital computer is not suited to performing this computation by way of calculus. The automatic programming of the fundamental theorem of the integral calculus on the digital computer is a formidable, if not impossible, task of providing the computer with a symbolic list of functions and the

1

corresponding antiderivatives (that is, a table of indefinite integrals), establishing a procedure for enlarging upon this list through combinations and substitutions, and having the computer decide when the integral can or cannot be expressed in terms of the listed functions. Moreover, even if the computer succeeds in expressing the integral in terms of these functions, and we easily imagine that it will often fail, a considerable amount of arithmetic still lies ahead since the integral, which can become vastly involved, need yet be evaluated at the limits of integration. Integral calculus is most potent, and hence its triumph in the precomputer days, in mental and hand computations of reasonable integrals that can be expressed in terms of common elementary functions. It is utterly unsuitable for automatic programming on the digital computer.

The way to compute the area A on the present-day computer is by the ancient precalculus technique of subdivision and summation. The domain whose area A we wish to compute is subdivided into small finite portions, the areas of which can be algebraically computed and summed together to provide an approximation for A. A better approximation is gained with a finer subdivision. The symbol manipulation of calculus, so alien to the digital computer, is avoided, and the computational procedure is reduced to an algebraic formulation for the sums from which a program is devised, instructing the computer to perform all of the many arithmetical operations that finally result in a *number* (or numbers) approximating A.

In our particular case, that of computing the area under $y = \cos x$, the interval $0 \leqslant x \leqslant \pi/2$ is subdivided into N, say equal, segments of finite length h, as shown in Fig. 1.1. A *lower bound* A_1 and an *upper bound* A_2 on A are computed by summing the areas of the rectangles and we have that

$$A_1 = h(\cos h + \cos 2h + \cdots + \cos Nh), \qquad Nh = \pi/2$$
$$A_2 = h(1 + \cos h + \cdots + \cos(N-1)h)$$

(1.2)

such that $A_1 \leqslant A \leqslant A_2$. Admittedly the algebraic formulations for A_1 and A_2 given in Eq. (1.2) are easy to program [assuming the computer knows $\cos(h)$], and the limiting process or *convergence* is carried out by increasing N—the number of subdivisions. The computer is incapable of performing the theoretical approach to the limit $N \to \infty$ but in practice this is also not essential. The area is desired only up to a certain practical accuracy that is mea-

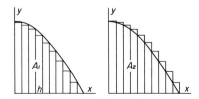

Fig. 1.1 A lower bound A_1 and an upper bound A_2 on the area under $y = \cos x$, $0 \leqslant x \leqslant \pi/2$.

sured by how close A_1 is to A_2, and a reasonable accuracy requirement is usually achieved on the computer quickly and at a fair price.

Equation (1.2) is specifically written for $y = \cos x$, but its generalization to any other curve is apparent. Moreover, this simple integration algorithm is infallible: It has nothing to do with the question of whether the integral is expressible in terms of some elementary functions or not. It might even be cheaper to compute the area by this algorithm when the integral is expressible in terms of tabulated functions, the values of which need be evaluated by another complicated program. Of course the practical execution of this program is feasible only with the fortunate help of the fast and accurate programmable computer. Without it, by hand calculation, the accurate numerical summation of the series (assuming no formula for the sum is found) for A_1 and A_2 may well become a formidable undertaking.

With differential equations the situation is similar. The theoretical limiting process in the mathematical formulation creates *differential* equations to describe physical phenomena or engineering processes. A symbolic solution to these equations in terms of some elementary functions, even if existing, is, as a rule, very hard to come by and the programming necessary to obtain such a symbolic solution impossible. To overcome this difficulty, numerical techniques of great generality have been invented, based on the *discretization* of the problem, on the division of the continuous flow of events or continuous change of state into a series of discrete states formulated algebraically, with the limiting process or convergence deferred to the numerical stage of the solution.

2. Differentials and Finite Difference Approximations

To fix ideas, let us consider in detail the formation and discretization of the differential equation describing the variation of atmospheric pressure with height. To this end we isolate a small cylindrical volume of air shown in Fig. 1.2 and apply to it the relevant laws of nature required to establish its equilibrium. As is commonly done by physicists, we *idealize* the situation in order to simplify the mathematical analysis, and still retain a valuable model. We thus

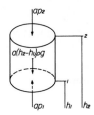

Fig. 1.2 A cylindrical element of air between heights h_1 and h_2, with hydrostatic and weight forces acting on it.

assume that the air is stationary, of constant temperature, obeys the laws of an ideal gas, and has properties that vary uniformly and continuously (uninterruptedly) with the height h. That we considerably simplify nature is obvious. Let the small imaginary air cylinder in Fig. 1.2 be located between the heights h_1 and h_2 and have a base area a. Only forces in the vertical direction need be considered since the pressure p and density ρ of the air are assumed to be uniform at any given height. These forces are $p_1 a$ exerted on the bottom of the cylinder by the pressure p_1 at that level, $p_2 a$ on top of the cylinder, and the weight $a(h_2 - h_1)\rho g$, in which ρ is the mean density and g the gravitational acceleration. Inasmuch as the cylinder of air is at rest, the algebraic sum of forces acting on it is zero resulting in the equation of equilibrium

$$p_1 a - p_2 a - a(h_2 - h_1)\rho g = 0 \tag{1.3}$$

or

$$(p_2 - p_1)/(h_2 - h_1) = -\rho g \tag{1.4}$$

Having assumed the air to be perfect gas, we have from Boyle's law that its density is proportional to the pressure, or $\rho = Rp$, R being the constant of proportionality, and Eq. (1.4) becomes

$$(p_2 - p_1)/(h_2 - h_1) = -Rgp \tag{1.5}$$

By our assumption that the pressure is a continuous function of h, we may proceed to the limit $h_2 - h_1 \to 0$, that occasions $p_2 - p_1 \to 0$ and we get the differential equation of atmospheric pressure

$$\frac{dp}{dh} = -Rgp \tag{1.6}$$

to which we add the initial condition $p = p_0$ at $h = 0$. There is nothing in Eq. (1.6) to fret about; we know well that the symbolic solution to this *initial value problem* is

$$p = p_0 e^{-Rgh} \tag{1.7}$$

Nevertheless, this simple problem serves well to illustrate the way the physical problem or differential equation is discretized and we retain it for that purpose.

One way to discretize the change of pressure with height is to halt the limiting process at Eq. (1.5) and consider the pressure only at discrete heights h. The pressure p appearing in Eq. (1.5) is average and we replace it *approximately* by $(p_1 + p_2)/2$, by which the equilibrium Eq. (1.5) becomes approximately

$$(p_2 - p_1)/(h_2 - h_1) = -\tfrac{1}{2}Rg(p_1 + p_2) \tag{1.8}$$

and consequently

$$p_2 = \frac{1 - \frac{1}{2}Rg(h_2 - h_1)}{1 + \frac{1}{2}Rg(h_2 - h_1)} p_1 \tag{1.9}$$

permitting us to compute the approximate pressure p_2 at level h_2 from the previous pressure p_1 at level h_1, starting with p_0 at $h = 0$. At level n we have that

$$p_n = z^n p_0, \qquad z = \frac{1 - \frac{1}{2}Rg(h_2 - h_1)}{1 + \frac{1}{2}Rg(h_2 - h_1)} \tag{1.10}$$

where z is termed the *magnification factor* for the pressure. This algebraic formulation for the approximate pressure at any level n is computer executed with ease and the pressure accuracy is increased by diminishing $h_2 - h_1$.

Otherwise, instead of discretizing the process directly from physical considerations, we could have used the differential equation (1.6) (which can be found in the mathematical literature) and approximated it by replacing the differentials dp and dh by the *finite differences* $h_2 - h_1$ and $p_2 - p_1$, respectively, and p by $(p_1 + p_2)/2$. This latter approach is often preferred since correct differential equations and boundary, or initial, conditions have been set up for all interesting physical phenomena and can be found in the relevant literature.

3. Finite Difference Schemes

Replacement of differentials by finite differences is the paramount step in the algebraization of the differential equation and we wish to pay close attention, in this section, to this act. An obvious way to replace the first derivative y' by finite differences is to approximate the tangent by the chord as in Fig. 1.3. At any point within the interval $x_1 \leqslant x \leqslant x_2$, y' is approximated by

$$y' = h^{-1}(y_2 - y_1) \tag{1.11}$$

Fig. 1.3 Orientation of chord and tangents on a smooth function y of x between $x = x_1$ and $x = x_2$.

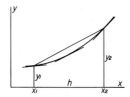

For most points in the interval this is only an approximation, but by Rolle's theorem there is at least one point inside the interval at which true equality holds. Equation (1.11) is, however, always exact for a linear y, where $y = a_0 + a_1 x$. In case the finite difference scheme in Eq. (1.11) is assigned to approximate y' at point 1 $(x = x_1)$, it is termed a *backward* scheme. It is a *forward* scheme if it is elected to approximate y' at point 2 $(x = x_2)$, and is a *central* scheme if assigned to approximate y' midway between 1 and 2, at $x = (x_1 + x_2)/2$. It is graphically hinted in Fig. 1.3 that the central scheme is more accurate than either the backward or forward scheme, the tangent at the midpoint seeming to be aligned more with the chord than the tangents at the end points. In the next section we shall prove this to be true.

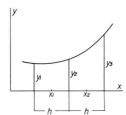

Fig. 1.4   Sampled values of y: y_1, y_2, and y_3 for the finite difference approximation of y''.

A finite difference scheme for y'' is generated, with reference to Fig. 1.4, by successively forming $(y')'$. Three values of y are needed for this, sampled at the three points 1, 2, 3, placed at a distance h apart. Let x_1 and x_2 be the midpoints between 1, 2 and 2, 3 in Fig. 1.4, and also let y_1' and y_2' denote the (approximate) value of y' at x_1 and x_2. Then from Eq. (1.11) we have that

$$y_1' = h^{-1}(y_2 - y_1), \qquad y_2' = h^{-1}(y_3 - y_2) \tag{1.12}$$

and therefore

$$y'' = h^{-1}(y_2' - y_1') \tag{1.13}$$

from which

$$y'' = h^{-2}(y_1 - 2y_2 + y_3) \tag{1.14}$$

emerges to approximate y'' in the interval between points 1 and 3. Distinction can again be made among backward, central, and forward difference schemes, depending on the point at which Eq. (1.14) is chosen to approximate y''. Equation (1.14) is only approximate except for $y = a_0 + a_1 x + a_2 x^2$, for which it is exact; when $y(x)$ is parabolic, the difference scheme of Eq. (1.14) finds its exact constant second derivative.

A more systematic way to construct finite difference schemes will emerge from the error analysis of the next section.

4. Accuracy Analysis

What errors are inherent in replacing differentials by finite differences and on what factors these errors depend is revealed by the following error analysis, in which Taylor's theorem plays the central role. First we undertake to assess the accuracy of the difference scheme for y' in Eq. (1.11) when used as a *central* scheme. For convenience we favor the origin of the coordinate system at the midpoint between 1 and 2, which we label 0 as in Fig. 1.5. No generality is lost in this choice since the accuracy of the finite difference scheme certainly depends on the intrinsic properties of y and the mesh size h, but is independent of the arbitrary location of the coordinate system.

Fig. 1.5 Central placement of coordinates to ease the error analysis of $(y_2 - y_1)/h$ approximating y_0'.

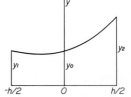

Using Taylor's theorem with remainder we expand y_1 and y_2 around $x = 0$ and have that

$$y_1 = y_0 - \frac{1}{2} h y_0' + \frac{1}{2!} \left(\frac{1}{2} h \right)^2 y_0'' - \frac{1}{3!} \left(\frac{1}{2} h \right)^3 y'''(\xi_1), \qquad -\frac{1}{2} h \leqslant \xi_1 \leqslant 0$$

$$\text{(1.15)}$$

$$y_2 = y_0 + \frac{1}{2} h y_0' + \frac{1}{2!} \left(\frac{1}{2} h \right)^2 y_0'' + \frac{1}{3!} \left(\frac{1}{2} h \right)^3 y'''(\xi_2), \qquad 0 \leqslant \xi_2 \leqslant \frac{1}{2} h$$

Subtraction of y_2 and y_1 in Eq. (1.15) yields

$$\frac{1}{h} (y_2 - y_1) = y_0' + \frac{1}{3!} \left(\frac{1}{2} \right)^3 h^2 [y'''(\xi_1) + y'''(\xi_2)] \qquad \text{(1.16)}$$

and therefore

$$\left| \frac{1}{h} (y_2 - y_1) - y_0' \right| \leqslant \frac{1}{24} h^2 \max|y'''(x)| \qquad \text{(1.17)}$$

If $y'''(x)$ is bounded between points 1 and 2, the error in replacing y_0' by the central difference scheme is bounded by a constant times h^2, or is *of order* h^2; in short $O(h^2)$. Halving h causes the bound on the error in this replacement to decrease by a factor of 4. We have already seen that there is no error in this replacement when y is linear. Equation (1.17) tells us that *at the center* there is no error in the finite difference approximation even if y is parabolic, for which $y'''(x) = 0$.

The inequality in Eq. (1.17) was created by replacing $|y'''(\xi_1) + y'''(\xi_2)|$ by the possibly larger $2 \max|y'''(x)|$. An inequality is termed *optimal* or *sharp* if equality does actually occur for some function y. Inequality (1.17) is sharp. Indeed for $y = x^3$, $y_0' = 0$, $y_1 = -h^3/8$, and $y_2 = h^3/8$ so that $(y_2 - y_1)/h - y_0' = h^2/4$. From inequality (1.17) we have, since $y''' = 6$, that $|(y_2 - y_1)/h - y_0'| \leq h^2/4$, giving exactly the same result.

The forward and backward schemes are less accurate. For the case in which $(y_2 - y_1)/h$ is taken to approximate y' at point 1, we expand y_2 in terms of the values of y at point 1 and have by Taylor's theorem that

$$y_2 = y_1 + hy_1' + \tfrac{1}{2}h^2 y''(\xi) \tag{1.18}$$

in which ξ is somewhere between points 1 and 2. Consequently

$$\left|\frac{1}{h}(y_2 - y_1) - y_1'\right| \leq \frac{1}{2} h \max|y''(x)| \tag{1.19}$$

where $\max| \ |$ stands for the maximum value between points 1 and 2, and the backward scheme is only $O(h)$. Halving the interval size h only halves the error. It is the same for the forward scheme.

The error analysis of the central difference scheme for y'' given in Eq. (1.14) is analogous. Again, we choose the origin at the central point 2 and expand y_1 and y_3 around point 2 to get by Taylor's theorem

$$y_1 = y_2 - hy_2' + \frac{1}{2!} h^2 y_2'' - \frac{1}{3!} h^3 y_2''' + \frac{1}{4!} h^4 y''''(\xi_1), \qquad -h \leq \xi_1 \leq 0 \tag{1.20}$$

$$y_3 = y_2 + hy_2' + \frac{1}{2!} h^2 y_2'' + \frac{1}{3!} h^3 y_2''' + \frac{1}{4!} h^4 y''''(\xi_2), \qquad 0 \leq \xi_2 \leq h$$

Addition of these two equations yields

$$\frac{1}{h^2}(y_1 + y_3 - 2y_2) = y_2'' + \frac{1}{4!} h^2 [y''''(\xi_1) + y''''(\xi_2)] \tag{1.21}$$

or

$$\left|\frac{1}{h^2}(y_1 - 2y_2 + y_3) - y_2''\right| \leq \frac{1}{12} h^2 \max|y''''(x)| \tag{1.22}$$

indicating that the central difference scheme is $O(h^2)$ if y'''' is finite. Once more the central scheme is more accurate than suspected. We realized that the central finite difference scheme of Eq. (1.14) is exact, at any point, for a parabolic y. According to Eq. (1.22), *at the center*, the difference scheme is also exact for a cubic y, for which $y'''' = 0$. It is readily verified that when the scheme in Eq. (1.14) is used in either the backward or forward sense, its accuracy drops from $O(h^2)$ to $O(h)$.

5. Higher-Order Schemes

In Section 3 two difference schemes, one for approximating y' and the other for y'', were created from geometrical considerations, and their accuracies were later analyzed in Section 4. Presently we wish to draw conclusions from that analysis that will point out to us a way to systematically generate more accurate, or higher-order, schemes. Let us reconsider the *central* difference scheme for y'' in Eq. (1.14), which we write now as

$$y_2'' = \frac{1}{h^2}(\alpha_1 y_1 + \alpha_2 y_2 + \alpha_3 y_3) \tag{1.23}$$

where α_1, α_2, and α_3 are parameters to be determined for highest accuracy. Expanding y_1 and y_3 around y_2 at $x = 0$, we get that

$$y_2'' = \frac{1}{h^2}\left[y_2(\alpha_1 + \alpha_2 + \alpha_3) + hy_2'(-\alpha_1 + \alpha_3) + \frac{1}{2!}h^2 y_2''(\alpha_1 + \alpha_3) \right.$$

$$\left. + \frac{1}{3!}h^3 y_2'''(-\alpha_1 + \alpha_3) + \frac{1}{4!}h^4 y_2''''(\alpha_1 + \alpha_3) + \cdots \right] \tag{1.24}$$

In order to make the right-hand side of Eq. (1.24) as close to y_2'' as possible when $h \to 0$, we set

$$\begin{aligned}
\alpha_1 + \alpha_2 + \alpha_3 &= 0 \\
-\alpha_1 + \alpha_3 &= 0 \\
\alpha_1 + \alpha_3 &= 2!
\end{aligned} \tag{1.25}$$

yielding $\alpha_1 = 1$, $\alpha_2 = -2$, and $\alpha_3 = 1$. Since the term with h^3 also disappears from the right-hand side of Eq. (1.24) with these values of α, the accuracy of this scheme becomes $O(h^2)$. The set of Eqs. (1.25) means that y_2'' in Eq. (1.23) is required to be exact for a constant, linear, and quadratic y. Indeed, when $y = 1$ and $y = x$, $y_2'' = 0$ and, consequently, the coefficients of y_2 and y_2' in Eq. (1.24) vanish. When $y = x^2$, $y_2 = 0$, $y_2' = 0$, $y_2'' = 2$, $y_2''' = 0$, etc., and hence $(\alpha_1 + \alpha_3)/2! = 1$.

Any other finite difference scheme may be generated in this way. The approximate derivative is written in terms of y sampled at some points as a weighted sum with weighting coefficients α, as in Eq. (1.23), to be determined by the condition that the scheme be exact for a polynomial y of highest possible degree.

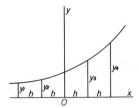

Fig. 1.6   Sampling points of y for a higher-order finite difference approximation of y'.

A higher-order central difference scheme for y' is set up, for instance, by sampling y at the four stations $x = -2h$, $x = -h$, $x = h$, and $x = 2h$, as shown in Fig. 1.6. The approximate y' at $x = 0$ is written as

$$y_0' = \alpha_1 y_1 + \alpha_2 y_2 + \alpha_3 y_3 + \alpha_4 y_4 \tag{1.26}$$

with α_1, α_2, α_3, and α_4 adjusted to provide highest accuracy for y' at $x = 0$. Having four parameters at our disposal, we ask y' in Eq. (1.26) to become exact for $y = 1$, $y = x$, $y = x^2$, and $y = x^3$, and get from this four equations

$$\begin{bmatrix} 1 & 1 & 1 & 1 \\ -2h & -h & h & 2h \\ 4h^2 & h^2 & h^2 & 4h^2 \\ -8h^3 & -h^3 & h^3 & 8h^3 \end{bmatrix} \begin{bmatrix} \alpha_1 \\ \alpha_2 \\ \alpha_3 \\ \alpha_4 \end{bmatrix} = \begin{bmatrix} 0 \\ 1 \\ 0 \\ 0 \end{bmatrix} \tag{1.27}$$

that upon solution yield $\alpha_1 + \alpha_4 = 0$, $\alpha_2 + \alpha_3 = 0$, $\alpha_1 = 1/(12h)$, and $\alpha_2 = -2/(3h)$. Now

$$y_0' = 1(12h)(y_1 - 8y_2 + 8y_3 - y_4) \tag{1.28}$$

which happens to be exact not only for $y = x^3$ but also for $y = x^4$. It is a property of the central difference schemes for y' that, because of symmetry, they are also exact for the next even power of x. The error in y' using Eq. (1.28), is proportional to $\max|y^{(5)}(x)|$, $-2h \leqslant x \leqslant 2h$, and is $O(h^4)$.

Higher-order finite difference schemes may be constructed not by sampling y at more stations, but rather by including higher-order derivatives of y in the scheme. A backward finite difference scheme of this form for y'' is written as

$$y_1'' = \alpha_1 y_1 + \beta_1 y_1' + \alpha_2 y_2 + \beta_2 y_2' \tag{1.29}$$

where $y_1 = y(-h)$ and $y_2 = y(h)$. Requiring that Eq. (1.29) be correct for $y = 1, x, x^2$, and x^3 at $x = -h$ yields the system of equations

$$\begin{bmatrix} 1 & & 1 & \\ -h & 1 & h & 1 \\ h^2 & -2h & h^2 & 2h \\ -h^3 & 3h^2 & h^3 & 3h^2 \end{bmatrix} \begin{bmatrix} \alpha_1 \\ \beta_1 \\ \alpha_2 \\ \beta_2 \end{bmatrix} = \begin{bmatrix} 0 \\ 0 \\ 2 \\ -6h \end{bmatrix} \tag{1.30}$$

solved by $\alpha_1 = -3/(2h^2)$, $\alpha_2 = 3/(2h^2)$, $\beta_1 = -2/h$, and $\beta_2 = -1/h$, and the scheme in Eq. (1.29) becomes

$$y_1'' = 3/(2h^2)(y_2 - y_1) - 1/(h)(2y_1' + y_2') \qquad (1.31)$$

To assess the error committed in using Eq. (1.31) we choose $y = \frac{1}{24}\alpha x^4$, and have that $y_1'' = \frac{1}{2}\alpha h^2$. The approximate y_1'' obtained from Eq. (1.31) with $y = \frac{1}{24}\alpha x^2$ is $y_1'' = \frac{1}{6}\alpha h^2$, and the error in y_1'' is proportional to α and is $O(h^2)$.

EXERCISES

1. Form a three-point forward, then backward, difference scheme for y' in the form $y' = \alpha_1 y_1 + \alpha_2 y_2 + \alpha_3 y_3$, where $y_1 = y(-h)$, $y_2 = y(0)$, $y_3 = y(h)$.

2. Estimate the accuracy of the scheme in Exercise 1.

3. Form a five point central difference scheme for y'' in the form $y_3'' = \alpha_1 y_1 + \alpha_2 y_2 + \alpha_3 y_3 + \alpha_4 y_4 + \alpha_5 y_5$. Choose the points 1, 2, 3, 4, and 5 to be equidistant.

4. Estimate the accuracy of the scheme in Exercise 3.

5. Use the right-hand side of Eq. (1.29) to form a central difference scheme for y''.

Suggested Further Reading

Goldberg, S., *Difference Equations*. Wiley, New York, 1967.

2 Two-Point Boundary Value Problems

1. Finite Difference Approximation of the Loaded String Equation

Having reviewed the approximate replacement of differentials by finite differences, we advance to apply this approximation to entire differential equations. A simple differential equation which is of physical significance, fitting for the illustration of finite difference discretization, is that of the stretched string, under unit tension, fixed at both ends, elastically supported on a flexible bed, and loaded along its length, as shown in Fig. 2.1. The string is assumed to be of length 2, situated between $x = 0$ and $x = 2$, of a variable spring constant $q(x) \geq 0$, and acted upon by a distributed load, or forcing function, $f(x)$, with both $q(x)$ and $f(x)$ symmetric about the center of the string at $x = 1$. Specifying the *small* deflection $y(x)$ of the string is the *linear* differential equation

$$-y'' + q(x)y = f(x), \qquad 0 < x < 2 \tag{2.1}$$

with boundary conditions

$$y(0) = y(2) = 0 \tag{2.2}$$

for the fixed ends of the string. Equations (2.1) and (2.2) constitute a linear *two-point boundary value problem*. Because of its symmetry, and that of the load, the string can be halved; the differential equation (2.1) considered

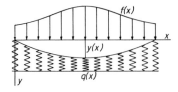

Fig. 2.1 Stretched string on an elastic bed of variable stiffness $q(x)$, and symmetrically loaded by the distributed force $f(x)$.

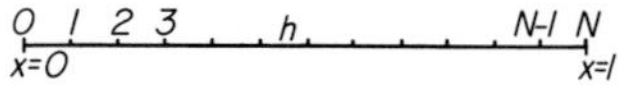

Fig. 2.2 The interval $0 \leqslant x \leqslant 1$ uniformly subdivided into segments of size h, and with the nodal points sequentially numbered $0, 1, \ldots, N - 1, N$.

only inside $0 < x < 1$ and the boundary condition taken to be

$$y(0) = 0, \qquad y'(1) = 0 \tag{2.3}$$

To discretize Eq. (2.1) with finite differences, the interval $0 < x < 1$ is divided into N conveniently equal sections, each of size $h = 1/N$, as shown in Fig. 2.2. The joints of these small sections are the *nodal points*, labeled sequentially $0, 1, 2, \ldots, N$. We designate the measured or computed values of y, q, and f by y_j, q_j, and f_j, respectively, at the jth node. Equation (2.1) holds true for any point x on the string. Its discretization is accomplished by replacing each term of the differential equation by finite difference approximations only at the nodes. To replace y'' we choose the accurate central scheme of Eq. (1.14), while we simply replace $f(x)$ and $q(x)y$ by f_j and $q_j y_j$, and we write for each *interior* nodal point from 1 to $N - 1$ the finite difference equation

At point 1: $\qquad h^{-2}(-y_0 + 2y_1 - y_2) + q_1 y_1 = f_1$

At point 2: $\qquad h^{-2}(-y_1 + 2y_2 - y_3) + q_2 y_2 = f_2 \tag{2.4}$

$\qquad \qquad \vdots$

At point $N - 1$: $\quad h^{-2}(-y_{N-2} + 2y_{N-1} - y_N) + q_{N-1} y_{N-1} = f_{N-1}$

In matrix form, Eq. (2.4) is organized into

$$\frac{1}{h^2} \left\{ \begin{bmatrix} -1 & 2 & -1 & & & \\ & -1 & 2 & -1 & & \\ & & \ddots & \ddots & \ddots & \\ & & & -1 & 2 & -1 \end{bmatrix} + \begin{bmatrix} q_1 & & & \\ & q_2 & & \\ & & \ddots & \\ & & & q_{N-1} \end{bmatrix} \right\} \begin{bmatrix} y_0 \\ y_1 \\ y_2 \\ \vdots \\ y_{N-1} \\ y_N \end{bmatrix}$$

$$= \begin{bmatrix} f_1 \\ f_2 \\ \vdots \\ f_{N-1} \end{bmatrix} \tag{2.5}$$

from which we hope to extract the nodal unknowns $y_0, y_1, \ldots, y_N$.

There are, however, $N + 1$ unknown ys in Eq. (2.5) but only $N - 1$ equations. The missing two equations are supplied by the two boundary conditions $y(0) = 0$ and $y'(1) = 0$.

2.   Incorporation of Boundary Conditions

Enforcement of the boundary condition $y(0) = 0$ is straightforward. At this end point $y_0 = 0$, which can either be added as the first equation in the system (2.5), or better, y_0 can be set equal to zero in the first of Eqs. (2.4) and y_0 excluded from among the entries of the list of unknowns. The advantage of this latter approach is that it works toward upholding *symmetry* in the first matrix of Eq. (2.5). Inclusion of the second boundary condition $y'(1) = 0$ at the string's point of symmetry is more subtle.

The first choice that occurs to us for the approximation of $y'(1)$ is the forward scheme

$$y'(1) = h^{-1}(y_N - y_{N-1}) \tag{2.6}$$

But this approximation has the accuracy of only $O(h)$, which we feel is out of balance with the $O(h^2)$ scheme used in the interior to approximate y''. A low accuracy approximation for $y'(1)$, we fear, may adversely affect the accuracy of the computed y at all points. An $O(h^2)$ central finite difference scheme for y' at nodal point N is thus desired, and can actually be secured through the addition of an $N + 1$ ficticious nodal point at $x = 1 + h$. With this extra point, point N becomes interior and the finite difference equation

$$h^{-2}(-y_{N-1} + 2y_N - y_{N+1}) + q_N y_N = f_N \tag{2.7}$$

may be written for it. For $y'(1)$ we now write the more accurate central approximation

$$y'(1) = (1/2h)(y_{N+1} - y_{N-1}) \tag{2.8}$$

Elimination of y_{N+1} between Eqs. (2.7) and (2.8) leaves us with

$$h^{-2}(-y_{N-1} + y_N) + \tfrac{1}{2}q_N y_N = \tfrac{1}{2}f_N + h^{-1}y'(1) \tag{2.9}$$

which is written, after setting $y'(1) = 0$, as the last equation in the system (2.5) that finally becomes

$$
\left\{ \frac{1}{h^2}
\begin{bmatrix}
2 & -1 & & & \\
-1 & 2 & -1 & & \\
 & \cdot & \cdot & \cdot & \\
 & & -1 & 2 & -1 \\
 & & & -1 & 1
\end{bmatrix}
+
\begin{bmatrix}
q_1 & & & & \\
 & q_2 & & & \\
 & & \cdot & & \\
 & & & q_{N-1} & \\
 & & & & \tfrac{1}{2}q_N
\end{bmatrix}
\right\}
\begin{bmatrix}
y_1 \\
y_2 \\
\vdots \\
y_{N-1} \\
y_N
\end{bmatrix}
$$

$$
=
\begin{bmatrix}
f_1 \\
f_2 \\
\vdots \\
f_{N-1} \\
\tfrac{1}{2}f_N
\end{bmatrix}
\tag{2.10}
$$

In short

$$Ky = f \tag{2.11}$$

in which K is an $N \times N$ symmetric matrix, y the vector containing the nodal unknowns, and f the forcing vector. *Finite differences effected the replacement of the linear boundary value problem—differential equation plus boundary conditions—by a system of linear algebraic equations that when solved furnishes approximate y values at the nodes.* The solvability of the linear algebraic system set up by finite differences has yet to be proved. Assuming for the moment the *numerical* existence of K^{-1} this system can, infallibly and at a reasonable cost, be solved on the computer by a variety of programmed direct or iterative techniques that have been developed for that purpose in the context of computational linear algebra.

Later on we shall prove the general nonsingularity of K for a wide class of boundary value problems discretized by *finite elements*. Presently we briefly notice some of the more obvious properties of K: It is symmetric, sparse, banded with bandwidth 3—which makes it tridiagonal—and of a repetitive structure. Since K is symmetric, the eigenvalues of K are all real. Gerschgorin's theorem asserts that all eigenvalues λ^K of K are confined to

$$\tfrac{1}{2}\min_j\{q_j\} \leqslant \lambda^K \leqslant 4h^{-2} + \max_j\{q_j\} \tag{2.12}$$

predicting that when $q > 0$, K is *positive definite*, all its eigenvalues are larger than $\tfrac{1}{2}q_{\min}$, and K^{-1} exists. Another interesting property of K, observed upon its inversion and related to the *maximum principle* in second-order boundary value problems, is that K^{-1} is *positive*, or $K_{ij}^{-1} > 0$ for all i and j. In contrast to the matrix K, which is sparse, the matrix K^{-1} is *dense*.

According to Eq. (2.12), the *spectral condition number* $C_2(K)$ of K, defined as

$$C_2(K) = \max_j |\lambda_j{}^K| \Big/ \min_j |\lambda_j{}^K| \tag{2.13}$$

is bounded by

$$C_2(K) \leqslant 8h^{-2}/q_{\min} + 2q_{\max}/q_{\min} \tag{2.14}$$

Possibly the most noticeable feature of the bound in Eq. (2.14) is the appearance of h^{-2} in it, predicting a decline in the condition of K, with a mesh refinement to achieve a higher discretization accuracy. This decline actually occurs, but nevertheless, for reasonable values of h, K is usually adequately well conditioned for its accurate inversion on existing computers.

When $q(x) = 0$, Gerschgorin's theorem fails to confirm the positive definiteness of K; but K still is positive definite. For the particular case of the K matrix in Eq. (2.10), when $q(x) = 0$, it is easy to find an analytic expression for all its eigenvalues and to verify that $C_2(K) = O(h^{-2})$. In more complex situations with higher-order schemes, pure linear algebraic arguments are extremely complicated (if not utterly powerless) generally to predict the condition of K made with finite differences and one has to revert to numerical methods. With finite elements the situation is different as we shall see when we come to Chapter 7.

3.   Consistency and Stability: Convergence

It does not escape us that the vector y in Eq. (2.10) contains only the *approximate* values of $y(x)$ at the nodes. Next, we shall assess the error $y_j - y(x)$ as depending on h and the properties of $y(x)$. A first step in this direction is the determination of the *local truncation error* r_j occurring in the difference equation when the *exact* y values at the jth node Y_j are introduced in the place of y_j. The residual, or imbalance, r_j caused in a typical finite difference equation (2.4) when Y_j is introduced into it instead of y_j is

$$r_j = h^{-2}(-Y_{j-1} + 2Y_j - Y_{j+1}) + q_j Y_j - f_j \qquad (2.15)$$

which for typographical brevity we prefer to write as

$$r_1 = h^{-2}(-Y_0 + 2Y_1 - Y_2) + q_1 Y_1 - f_1 \qquad (2.16)$$

Expansion of Y_0 and Y_2 in a Taylor series around point 1 results in

$$r_1 = (-Y_1'' + q_1 Y_1 - f_1) - \frac{1}{4!} h^2 [y''''(\xi_1) + y''''(\xi_2)] \qquad (2.17)$$

where ξ_1 and ξ_2 are somewhere between points 0 and 2. Inasmuch as Y_1 is the exact value of $y(x)$ at point 1, Y_1 satisfies the differential equation, the expression enclosed in parentheses in Eq. (2.17) vanishes, and $r_1 = O(h^2)$, provided $y''''(x) <$ const. between points 0 and 2. In the event that $r_j \to 0$ as $h \to 0$, the finite difference equation is said to be *consistent* with the original differential equation. Here consistency is $O(h^2)$. Consistency is a prerequisite for the convergence of y_j to Y_j as $h \to 0$, but it is not sufficient. The finite difference scheme must also be *stable* to ensure convergence. We shall discuss the stability question shortly after having checked the consistency of the finite difference boundary approximation.

The last of Eqs. (2.10), corresponding to the boundary condition $y'(1) = 0$, is different from the interior equations and we wish to consider its consistency separately. By Taylor's expansion we have that

$$Y_{N-1} = Y_N - h Y_N' + \frac{1}{2!} h^2 Y_N'' - \frac{1}{3!} h^3 Y_N''' + \frac{1}{4!} h^4 y''''(\xi) \qquad (2.18)$$

in which ξ is between points $N - 1$ and N. Since N (or $x = 1$) is a point of symmetry, $Y_N' = 0$ and $Y_N'' = 0$ at this point, and Eq. (2.18) is reduced to

$$Y_{N-1} - Y_N = \frac{1}{2!} h^2 Y_N'' + \frac{1}{4!} h^4 y''''(\xi) \tag{2.19}$$

Consequently, for the last of the finite difference equations (2.10)

$$r_N = \frac{1}{2} (- Y_N'' + q_N Y_N - f_N) - \frac{1}{4!} h^2 y''''(\xi) \tag{2.20}$$

and also at point N, $r_N = O(h^2)$.

To relate the truncation errors to the errors in y, we collect the exact nodal values in the vector Y, the computed values in y, the residuals at all points in r, and have from Eq. (2.11) concisely that

$$KY = f + r, \qquad Ky = f \tag{2.21}$$

Hence

$$K(Y - y) = r, \qquad Y - y = K^{-1}r \tag{2.22}$$

which can be given the following physical interpretation: *The displacement errors $Y_j - y_j$ are the discrete displacements of the string caused by a forcing vector r.* Now we realize that for the convergence of the displacements we need K^{-1} with the property that a small load will always cause a small deflection in the spring such that $r_j \to 0$ will occasion $Y_j - y_j \to 0$. More precisely put, since

$$\|Y - y\|_\infty \leq \|K^{-1}\|_\infty \|r\|_\infty \tag{2.23}$$

we first of all wish $\|K^{-1}\|_\infty$ to be finite, or K^{-1} computable, for any mesh size h. It may happen that $\|K^{-1}\|_\infty$ grows as h is decreased, but then this growth has to be sufficiently slow for $\|K\|_\infty \|r\|_\infty$ to *diminish* as $h \to 0$. This certainly happens when $\|K^{-1}\|_\infty$ is bounded from above by some constant independent of h. Then the finite difference scheme is said to be *stable*, and if the scheme is consistent and stable, obviously $\|Y - y\|_\infty \to 0$ as $h \to 0$; convergence of the computed displacements takes place. In fact, if $\|K^{-1}\|_\infty <$ const. independently of h, then for y obtained from Eq. (2.10) $\|Y - y\|_\infty \leq O(h^2)$.

Equation (2.22) points out that since K^{-1} is dense, the truncation error at any point j contributes to the error in the displacements at all other points.

Stability is difficult to establish in general for finite difference schemes but we are not troubled by this now since the finite element method we have in mind to discuss next for the generation of finite difference equations by variational means will give us a comprehensive, decisive, and favorable answer to the general question of stability.

4. Higher-Order Consistency

The central difference scheme used to replace y'' in the string equation was $O(h^2)$, and this happened also to be the order of the truncation errors, and eventually that of $\|Y - y\|_\infty$. It is shown next that the $O(h^2)$ in the difference scheme for y'' does not limit the order of the residuals to h^2, and that a higher-order truncation error can be realized for the string equation with more terms in the approximation of $q(x)y$ and $f(x)$. To observe this let us simplify the string equation to $-y'' + y = f(x)$. We propose to write the finite difference approximation to this equation, at a typical interior point 1, in the general form

$$h^{-2}(-y_0 + 2y_1 - y_2) + (\alpha_0 y_0 + \alpha_1 y_1 + \alpha_0 y_2) = (\beta_0 f_0 + \beta_1 f_1 + \beta_0 f_2) \quad (2.24)$$

and seek to adjust $\alpha_0, \alpha_1, \beta_0$, and β_1 for highest order of consistency. To fix the αs and βs, we replace y_j by Y_j and expand Y_0, Y_2, f_0, and f_2 around point 1, getting

$$\frac{1}{h^2}(-Y_0 + 2Y_1 - Y_2) = -Y_1'' - \frac{2}{4!}h^2 Y_1'''' - \frac{2}{6!}h^4 Y_1^{(6)} - \cdots$$

$$\beta_0 f_0 + \beta_1 f_1 + \beta_0 f_2 = f_1(2\beta_0 + \beta_1) + \frac{2}{2!}h^2 f_1'' \beta_0 + \frac{2}{4!}h^4 f_1'''' \beta_0 + \cdots \quad (2.25)$$

$$\alpha_0 Y_0 + \alpha_1 Y_1 + \alpha_0 Y_2 = Y_1(2\alpha_0 + \alpha_1) + \frac{2}{2!}h^2 Y_1'' \alpha_0 + \frac{2}{4!}h^4 Y_1'''' \alpha_0 + \cdots$$

which, when substituted in Eq. (2.24) yields

$$r_1 = [-Y_1'' + Y_1(2\alpha_0 + \alpha_1) - f_1(2\beta_0 + \beta_1)]$$
$$+ \frac{2}{4!}h^2[-Y_1'''' + 12Y_1'' \alpha_0 - 12\beta_0 f_1''] + O(h^4) \quad (2.26)$$

The choice of $2\alpha_0 + \alpha_1 = 1, 2\beta_0 + \beta_1 = 1, \alpha_0 = \frac{1}{12}, \beta_0 = \frac{1}{12}$ brings Eq. (2.26) into the form

$$r_1 = (-Y_1'' + Y_1 - f) + \tfrac{1}{12}h^2(-Y_1'''' + Y_1'' - f_1'') + O(h^4) \quad (2.27)$$

and since both $-Y_1'' + Y_1 - f_1$ and $-Y_1'''' + Y_1'' - f_1''$ equal zero, by virtue of Y_1 satisfying the differential equation $-y'' + y - f = 0, r_1 = O(h^4)$. Thus

the difference equation

$$h^{-2}(-y_0 + 2y_1 - y_2) + \tfrac{1}{12}(y_0 + 10y_1 + y_2) = \tfrac{1}{12}(f_0 + 10f_1 + f_2) \quad (2.28)$$

is consistent, $O(h^4)$, with the differential equation $-y'' + y = f(x)$. Its matrix K is such that $\|K^{-1}\|_\infty \leq$ const., and hence the accuracy of y computed from Eq. (2.28) is also, according to the analysis of the previous section, $O(h^4)$; with the same mesh size h the scheme in Eq. (2.28) is expected to furnish results of greater accuracy than those gotten from Eq. (2.4).

5.   Finite Difference Approximation of the Beam Equation

The differential equation describing the deflection of a stretched string includes at most y'' and is therefore of second order. That describing the deflection y of a laterally loaded thin beam includes y'''' and is of order four. In particular, for a beam of unit length, elastically supported on an elastic bed of variable stiffness $q(x) \geq 0$, and loaded by a distributed force $f(x)$, it is

$$(p(x)y'')'' + q(x)y = f(x), \qquad 0 < x < 1 \tag{2.29}$$

in which $p(x) > 0$ includes the elastic and geometric properties of the beam's cross section. We wish to assume a beam with the commonly occurring homogeneous boundary conditions

$$y(0) = 0, \qquad y''(0) = 0$$
$$y(1) = 0, \qquad y'(1) = 0 \tag{2.30}$$

At the end point $x = 0$ the beam is *hinged* or *simply supported*, while at the end point $x = 1$ it is *clamped* or *built-in* as shown in Fig. 2.3.

Fig. 2.3 Thin elastic beam simply supported at one end and clamped at the other.

At least five y values are needed in a finite difference scheme for y'''', which can be generated by successive applications of the finite difference formula for y'' at the points 2, 3, and 4 shown in Fig. 2.4. With points 2, 3, and 4

Fig. 2.4   Sampling points of y for a finite difference approximation of y''''.

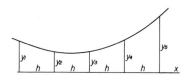

central, the difference schemes for y_2'', y_3'', and y_3'' become

$$y_2'' = h^{-2}(y_1 - 2y_2 + y_3), \qquad y_3'' = h^{-2}(y_2 - 2y_3 + y_4),$$
$$y_4'' = h^{-2}(y_3 - 2y_4 + y_5) \tag{2.31}$$

and since at point 3

$$y_3'''' = h^{-2}(y_2'' - 2y_3'' + y_4'') \tag{2.32}$$

the scheme

$$y_3'''' = h^{-4}(y_1 - 4y_2 + 6y_3 - 4y_4 + y_5) \tag{2.33}$$

which is $O(h^2)$, results for the approximation of $y''''(x)$. It strikes us that the coefficients in the central scheme for $y^{(n)}$ are those of the binomial $(1 - 1)^n$. This is not surprising in view of the fact that the scheme for $y^{(n)}$ is formed from n successive applications of the scheme for y_2', $y_2' = (y_3 - y_1)/2h$.

For $(p(x)y'')''$ the scheme in Eq. (2.33) becomes more involved:

$$(py'')'' = \frac{1}{h^4}\left[y_1 p_2 - 4y_2 \frac{p_2 + p_3}{2} + 6y_3 \frac{p_2 + 4p_3 + p_4}{6} - 4y_4 \frac{p_3 + p_4}{2} + y_5 p_4 \right] \tag{2.34}$$

To keep matters as simple as possible, we proceed with the finite difference discretization of the simpler beam equation in which $p(x) = 1$ and $q(x) = 0$.

Again, to discretize the boundary value problem in Eqs. (2.29) and (2.30), the beam is divided into N segments, equal for convenience, of size $h = 1/N$ and the nodal points are sequentially labeled from 0 to N. Equation (2.33) is then used to write the finite difference equations approximating $y'''' = f(x)$ at all interior points from 2 to $N - 2$, with four additional equations furnished by the four boundary conditions in Eq. (2.30). The two boundary conditions $y(0) = 0$ and $y(1) = 0$ are readily enforced by setting $y_0 = 0$ and $y_N = 0$, and removing both y_0 and y_N from the list of unknowns. To incorporate the other two boundary conditions $y''(0) = 0$ and $y'(1) = 0$ with an $O(h^2)$ central difference scheme, we employ the trick of additional fictitious nodal points at $x = -h$ and $x = 1 + h$, labeled -1 and $N + 1$. Now, at points 0 and N

$$y_0'' = (1/h^2)(y_{-1} - 2y_0 + y_1), \qquad y_N' = (1/2h)(y_{N+1} - y_{N-1}) \tag{2.35}$$

while the finite difference approximation of the differential equation at points 1 and $N - 1$ becomes

$$h^{-4}(y_{-1} - 4y_0 + 6y_1 - 4y_2 + y_3) = f_1$$
$$h^{-4}(y_{N-3} - 4y_{N-2} + 6y_{N-1} - 4y_N + y_{N+1}) = f_{N-1} \tag{2.36}$$

Elimination of y_{-1} and y_{N+1} between Eqs. (2.35) and (2.36) results, with the interior equations, in the linear algebraic system

$$
\frac{1}{h^4}
\begin{bmatrix}
5 & -4 & 1 \\
-4 & 6 & -4 & 1 \\
1 & -4 & 6 & -4 & 1 \\
 & 1 & -4 & 6 & -4 & 1 \\
 & & \ddots & \ddots & \ddots & \ddots & \ddots \\
 & & & 1 & -4 & 6 & -4 & 1 \\
 & & & & 1 & -4 & 6 & -4 & 1 \\
 & & & & & 1 & -4 & 6 & -4 \\
 & & & & & & 1 & -4 & 7
\end{bmatrix}
\begin{bmatrix}
y_1 \\ y_2 \\ y_3 \\ y_4 \\ \vdots \\ y_{N-4} \\ y_{N-3} \\ y_{N-2} \\ y_{N-1}
\end{bmatrix}
=
\begin{bmatrix}
f_1 \\ f_2 \\ f_3 \\ f_4 \\ \vdots \\ f_{N-4} \\ f_{N-3} \\ f_{N-2} \\ f_{N-1}
\end{bmatrix}
$$

$$(2.37)$$

In short $Ky = f$, with the matrix K symmetric, sparse, and banded, with bandwidth 5. Gerschgorin's theorem fails to predict the positive definiteness of K; but K is positive definite with spectral condition number $O(h^{-4})$. Many routines are available for the computer solution of linear algebraic systems with symmetric, banded, and positive definite matrices like the one in Eq. (2.37).

6. Splitting of the Beam Equation into Two String Equations

If one wishes to do so one may split the fourth-order beam equation (2.29) into the two second-order equations

$$-p(x)y'' = M, \qquad -M'' + q(x)y = f(x), \qquad 0 < x < 1 \qquad (2.38)$$

where the letter M is chosen to denote $-p(x)y''$ to signify that it is the *bending moment* in the beam's cross section. Simplicity compels us here too to choose $p(x) = 1$, $q(x) = 0$, but we retain the boundary conditions $y(0) = 0$, $y''(0) = 0$ ($M(0) = 0$), $y(1) = 0$, and $y'(1) = 0$.

To discretize the pair of Eqs. (2.38) the beam is divided into N parts and the nodes labeled from 0 to N. But now there are twice as many nodal values since each node is associated with both a y and an M value. These may be differently grouped to profoundly influence the nonzero entry distribution inside the K matrix. In case all Ms are considered first and the ys next, the corresponding matrix K becomes

$$
K = \frac{1}{h^2}
\left[
\begin{array}{ccccccc|ccccccc}
-h^2 & & & & & & & 2 & -1 & & & & & \\
& -h^2 & & & & & & -1 & 2 & -1 & & & & \\
& & -h^2 & & & & & & -1 & 2 & -1 & & & \\
& & & -h^2 & & & & & & -1 & 2 & -1 & & \\
& & & & -h^2 & & & & & & -1 & 2 & -1 & \\
& & & & & -h^2 & & & & & & -1 & 2 & -1 \\
& & & & & & -h^2 & & & & & & -1 & 2 \\
& & & & & & -\tfrac{1}{2}h^2 & & & & & & & -1 \\
\hline
2 & -1 & & & & & & & & & & & & \\
-1 & 2 & -1 & & & & & & & & & & & \\
& -1 & 2 & -1 & & & & & & & & & & \\
& & -1 & 2 & -1 & & & & & & & & & \\
& & & -1 & 2 & -1 & & & & & & & & \\
& & & & -1 & 2 & -1 & & & & & & & \\
\end{array}
\right]
\tag{2.39}
$$

where the equation with the $-h^2/2$ on the diagonal corresponds to the boundary condition $y'(1) = 0$. In case the unknowns are grouped in the pairs $M_1, y_1, M_2, y_2, \ldots$, the matrix K becomes

$$
K = \frac{1}{h^2}
\left[
\begin{array}{ccccccccccccc}
-h^2 & 2 & & -1 & & & & & & & & & \\
2 & & -1 & & & & & & & & & & \\
-1 & -h^2 & 2 & & -1 & & & & & & & & \\
-1 & & 2 & & -1 & & & & & & & & \\
& & -1 & -h^2 & 2 & & -1 & & & & & & \\
& & -1 & & 2 & & -1 & & & & & & \\
& & & & -1 & -h^2 & 2 & & -1 & & & & \\
& & & & -1 & & 2 & & -1 & & & & \\
& & & & & & -1 & -h^2 & 2 & & -1 & & \\
& & & & & & -1 & & 2 & & -1 & & \\
& & & & & & & & -1 & -h^2 & 2 & & -1 \\
& & & & & & & & -1 & & 2 & & -1 \\
& & & & & & & & & & -1 & -h^2 & 2 \\
& & & & & & & & & & -1 & & 2 & -1 \\
& & & & & & & & & & & & -1 & -\tfrac{1}{2}h^2 \\
\end{array}
\right]
\tag{2.40}
$$

with the last row corresponding to the boundary condition $y'(1) = 0$. Both matrices are symmetric, sparse, and banded, but the band in Eq. (2.40) is considerably narrower.

The numerical exploration of the other properties of K is left to the exercises. Here K is no longer positive definite; it is rather indefinite with negative and positive eigenvalues, a property of K that is detrimental to some solution algorithms for linear algebraic systems. On the other hand, its spectral condition number is only $O(h^{-2})$, which means that for the same number of divisions, K in Eq. (2.40) is better conditioned than its counterpart in Eq. (2.37). Also with splitting the discrete values of the bending moments are included in the solution, while in the technique of the previous section y'' needed to be computed separately from the y values calculated by the numerical solution procedure.

7. Nonlinear Two-Point Boundary Value Problems

Approximation of a nonlinear differential equation by finite differences is in principle not different from that of the previously discussed linear approximations, except that the resulting system of equations from this operation is *nonlinear*. While the solution of a linear system of equations is, at least in theory, routine, the solution of nonlinear equations is far from it. The solution of nonlinear systems is algorithmically more complicated, more time consuming, and the outcome less certain. Direct methods of solution are out of the question for nonlinear systems and recourse has to be made to iterative techniques.

The method of *successive substitutions* for the numerical solution of the nonlinear boundary value problem is one of the more attractive algorithms known: It is conceptually easy to program and is likely to converge if the nonlinearity is weak or if the solution to the boundary value problem is insensitive to the nonlinear terms.

Let us examine successive substitutions in the nonlinear equation

$$-y''(1 + \alpha y'^2) = f(x), \qquad y(0) = y(1) = 0 \qquad (2.41)$$

which, when $\alpha y'^2$ is negligible compared to 1 reduces to the linear string equation. As we shall see, the dependence of y on the nonlinear term is slight (hence the success of the linearization) and successive substitution is worth trying. In preparation we rewrite Eq. (2.41) in the form

$$-y'' = f(x)/(1 + \alpha y'^2) = F(x, y') \qquad (2.42)$$

and suggest the solution of Eq. (2.42) by the iterative scheme

$$-y''^{(n)} = F(x, y'^{(n-1)}) \qquad (2.43)$$

starting with $y^{(0)} = 0$. The first solution $y^{(1)}$ is obtained from the linear equation $-y'' = f(x)$, and is then used to correct the right-hand side of Eq. (2.42), and so on. When $y^{(n)}$ is sufficiently close to $y^{(n-1)}$, the iterative cycle is terminated. Because of the way Eq. (2.43) is set up, the matrix for it is as for the linear equation and does not change during the corrections of $y^{(n)}$. Only the forcing vector is different, and is here for $f(x) = 1$

$$f_1 = 1 \Big/ \left[1 + \alpha \left(\frac{y_2 - y_0}{2h} \right)^2 \right], \qquad y_0 = 0$$

$$f_2 = 1 \Big/ \left[1 + \alpha \left(\frac{y_3 - y_1}{2h} \right)^2 \right]$$

$$\vdots \tag{2.44}$$

$$f_{N-1} = 1 \Big/ \left[1 + \alpha \left(\frac{y_N - y_{N-2}}{2h} \right)^2 \right], \qquad y_N = 0$$

The successive corrections are shown in Fig. 2.5 for $\alpha = 4$ and $h = \frac{1}{12}$. Convergence occurs at $n = 3$; entailing the solution of a system of linear equations three times.

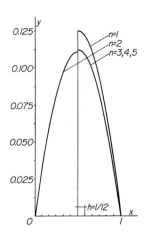

Fig. 2.5 Convergence of successive substitutions in $-y''^{(n+1)} = 1/(1 + 4y'^{(n)}y'^{(n)})$, starting with $y^{(0)} = 0$.

EXERCISES

1. The temperature distribution y within a unit disk with a central hole of radius $r = \frac{1}{2}$ is governed by the differential equation

$$-\frac{1}{r}\frac{d}{dr}\left(r\frac{dy}{dr} \right) + y = 0, \qquad \frac{1}{2} < r < 1$$

The disk is held at temperature 1 at $r = \frac{1}{2}$ and radiates heat at $r = 1$ such that the boundary conditions for the disk become

$$y = 1 \quad \text{at} \quad r = \frac{1}{2}, \qquad \frac{dy}{dr} + y = 1 \quad \text{at} \quad r = 1$$

Use finite differences to discretize this two point boundary value problem.

2. A unit elastic spherical shell of inner radius $r = \frac{1}{2}$ contains gas at pressure p. The differential equation describing the radial displacement y of the shell under the action of the internal pressure is

$$\frac{d}{dr}\left(\frac{dy}{dr} + \frac{2y}{r}\right) = 0$$

with boundary conditions

$$\frac{dy}{dr} = -p \quad \text{at} \quad r = \frac{1}{2}, \qquad \frac{dy}{dr} = 0 \quad \text{at} \quad r = 1$$

Use finite differences to discretize this problem.

3. The differential equation describing the lateral deflection y of a clamped unit circular plate point loaded is

$$\frac{1}{r}\frac{d}{dr}\left(r\frac{d}{dr}\left(\frac{1}{r}\frac{d}{dr}\left(r\frac{dy}{dr}\right)\right)\right) = P(0)$$

where $P(0)$ is a point force at $r = 0$, with boundary conditions

$$\frac{dy}{dr} = 0, \qquad y = 0 \qquad \text{at} \quad r = 1$$

Discuss the conditions at $r = 0$. Numerically solve this boundary value problem.

4. Discretize the equation $-y'' + y = 1$, $y(0) = y(1) = 0$, with the schemes in Eqs. (2.4) and (2.28). Make a computer study of the accuracy.

5. Compute the eigenvalues of K in Eq. (2.40) to observe that K is indefinite. Follow the change of its eigenvalues with h.

6. Use finite differences to discretize

$$-y'' + y\left(1 + \frac{1}{2}y^2\right) = 1, \qquad 0 < x < 1$$

$$y(0) = y(1) = 0$$

and solve this equation by successive substitutions, starting with

$$-y''^{(1)} + y^{(1)} = 1$$

and correcting according to

$$-y''^{(2)} + y^{(2)}(1 + \tfrac{1}{2}y^{(1)}y^{(1)}) = 1$$

7. Solve the nonlinear heat conduction boundary value problem

$$(e^{-\alpha y}y')' = 0, \qquad 0 < x < 1, \qquad y(0) = 0, \qquad y(1) = 1$$

for the values $\alpha = 0.1$, 0.5, and 1. Use the iterative scheme

$$(e^{-\alpha y^{(1)}}y'^{(2)})' = 0$$

starting with $y^{(1)} = 0$.

8. Consider the nonlinear equation

$$-y'' = f(y)$$

which can be written, for a moment, as

$$-y''^{(2)} = f(y^{(2)})$$

Expansion of $f(y^{(2)})$ around some previous $y^{(1)}$ is of the form

$$f(y^{(2)}) = f(y^{(1)}) + (y^{(2)} - y^{(1)}) \frac{\partial f(y^{(1)})}{\partial y}$$

suggesting the iterative scheme

$$-y''^{(2)} = f(y^{(1)}) + (y^{(2)} - y^{(1)})f'(y^{(1)})$$

Use this to solve

$$y'' = e^y, \qquad y(0) = y(1) = 0$$

and compare with the exact solution

$$y(x) = -\log 2 + 2\log\{c \sec 0.5c(x - 0.5)\}$$

where c is obtained from

$$\sqrt{2} = c \sec(0.25c)$$

or approximately $c = 1.3360557$.

9. Numerically solve the boundary value problem

$$-y'' = 1 + \alpha^2 y'^2, \qquad y(0) = y(1) = 0$$

and compare your result with the symbolic solution

$$y = \frac{1}{\alpha^2} \log\{\cos \alpha(x - 0.5)/\cos 0.5\alpha\}$$

10. A nonlinear differential equation of the form $-y'' = f(y, y')$ may be written as

$$-y''^{(2)} = f(y^{(2)}, y'^{(2)})$$

and the right-hand side expanded by

$$-y''^{(2)} = f(y^{(1)}, y'^{(1)}) + (y^{(2)} - y^{(1)})\frac{\partial f}{\partial y}(y^{(1)}, y'^{(1)})$$

$$+ (y'^{(2)} - y'^{(1)})\frac{\partial f}{\partial y'}(y^{(1)}, y'^{(1)})$$

to provide an iterative method for its solution. Use this procedure to solve the boundary value problem

$$yy'' + y'^2 + 1 = 0, \qquad y(1) = 1, \qquad y(2) = 2$$

that describes the passage of a ray of light through an optically inhomogeneous medium from point $(1, 1)$ to $(2, 2)$ in the xy plane. Show that in this case

$$-y''^{(2)} = \frac{1 + y'^{(1)}y'^{(1)}}{y^{(1)}} - \frac{1 + y'^{(1)}y'^{(1)}}{y^{(1)}y^{(1)}}(y^{(2)} - y^{(1)}) + \frac{2y'^{(1)}}{y^{(1)}}(y'^{(2)} - y'^{(1)})$$

Start your iterations with $y^{(1)} = x$ and compare your result with the exact solution $(x - 3)^2 + y^2 = 5$.

11. An elastic thin walled cylinder is held rigidly at its two ends and is under internal gas pressure $f(x)$. The equation describing its displacement is

$$\frac{t^2}{R^2}y'''' + y = f(x), \qquad 0 < x < 1$$

$$y(0) = y'(0) = y(1) = y'(1) = 0$$

in which t is the wall thickness and R the radius of this cylindrical shell. Solve this equation numerically, trace y, y', y'', and y''', and show that when $t/R \ll 1$ the bending effect introduced by y'''' in the equation is restricted to only a narrow *boundary layer* near the ends.

12. The displacement y of a laterally loaded stretched beam (stiff string) is given by an equation of the form

$$\varepsilon^2 y'''' - y'' = 1, \qquad 0 < x < 1$$
$$y(0) = y''(0) = y(1) = y''(1) = 0$$

Solve this problem numerically and observe the effect of $\varepsilon^2 \to 0$ on y, y', y'', and y'''.

13. Determine the accuracy of the central finite difference scheme

$$y_3'' = \frac{1}{12h^2}[4(y_1 - y_2 - y_4 + y_5) + \alpha(y_1 - 4y_2 + 6y_3 - 4y_4 + y_5)]$$

where the points $1-5$ are at equal distance h and where α is a parameter.

Find α that makes this scheme most accurate. Observe the schemes that result from $\alpha = -4$ and $\alpha = -1$. When the above scheme for y_3'' is used to discretize

$$-y'' = f(x) \qquad 0 < x < 1,\ y(0) = y(1) = 0$$

[Eq. (1.14) is used at the boundary points with fictious points added at $x = -h$ and $x = 1 + h$], the resulting global matrix is

$$K = \frac{1}{12h^2}\left\{ 4\begin{bmatrix} -1 & -1 & 1 & & & \\ -1 & & -1 & 1 & & \\ 1 & -1 & & -1 & 1 & \\ & 1 & -1 & & -1 & 1 \\ & & 1 & -1 & & -1 \\ & & & 1 & -1 & -1 \end{bmatrix} + \alpha \begin{bmatrix} 5 & -4 & 1 & & & \\ -4 & 6 & -4 & 1 & & \\ 1 & -4 & 6 & -4 & 1 & \\ & 1 & -4 & 6 & -4 & 1 \\ & & 1 & -4 & 6 & -4 \\ & & & 1 & -4 & 5 \end{bmatrix} \right\}$$

Use the computer to show that when $\alpha > -1$, K may become indefinite, some of its eigenvalues are negative, and that it is possible to find a value for α that makes K singular.

SUGGESTED FURTHER READING

Bennett, A. A., Milne, W. E., and Bateman, H., *Numerical Integration of Differential Equations*. Dover, New York, 1956.

Collatz, L., *The Numerical Treatment of Differential Equations*. Springer–Verlag, Berlin, 1960.

Levy, H., and Baggot E., *Numerical Solutions of Differential Equations*. Dover, New York, 1962.

Henrici, P., *Discrete Variable Methods in Ordinary Differential Equations*. Wiley, New York, 1962.

Henrici, P., *Error Propagation for Difference Methods*. Wiley, New York, 1963.

Godunov, S. K., and Ryabenki, V. S., *Theory of Difference Schemes*. North–Holland Publ., Amsterdam, 1964.

Bellman, R. E., and Kalaba, R. E., *Quasi Linearization and Nonlinear Boundary-Value Problems*. Amer. Elsevier, New York, 1965.

Greenspan, D., *Numerical Solutions of Nonlinear Differential Equations*. Wiley, New York, 1966.

Hildebrand, F. B., *Finite-Difference Equations and Simulation*, Prentice–Hall, Englewood Cliffs, New Jersey, 1968.

Keller, H. B., *Numerical Methods for Two-Point Boundary-Value Problems*. Ginn (Blaisdell), Boston, Massachusetts, 1968.

Bailey, P. B., Shampine, L. F., and Waltman, P. E., *Nonlinear Two Point Boundary Value Problems*. Academic Press, New York, 1968.

Daniel, J. W., and Moore, R. E., *Computation and Theory in Ordinary Differential Equations*. Freeman, San Francisco, California, 1970.

Milne, W. E., *Numerical Solution of Differential Equations*. Dover, New York, 1970.

Spiegel, M. R., *Finite Differences and Difference Equations*. McGraw-Hill, New York, 1971.

3 *Variational Formulations*

1. Energy Error

The finite difference solution of the previous chapter consisted of a list of values approximating y at the nodes only. The correctness of this approximation was inspected only at the nodes and it was measured in terms of the l_∞ vector norm. No attention was paid to the intervals between the nodal points. Next, we contemplate introducing approximations that cover all points within the range we are interested in. Appropriate error measures are then brought in for these approximations.

Figure 3.1 shows various functions $\tilde{y}$, candidates for the approximation of the continuous function y. Let $e(x) = y(x) - \tilde{y}(x)$ be the error distribution along x resulting from the approximation. As in vectors, also here, we wish to associate with $e(x)$ *one single positive number* to measure its magnitude. If this number is small, we shall say that the approximation is good; if large, we shall

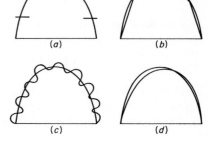

Fig. 3.1 Various approximations $\tilde{y}$ to a smooth function y.

say that the approximation is bad. This is the *norm* of e and is denoted by $\|e(x)\|$. Vertical lines are used since they remind us of the absolute value of a number and two since e is a function. An obvious choice for $\|e\|$ is the L_∞ norm

$$\|e\|_\infty = \max_x |e(x)| \qquad (3.1)$$

which is meaningful for any *finite* function. It is an acceptable error measure, or norm, by the fact that it satisfies the three fundamental conditions on the norm

$$\|e\| \geqslant 0$$
$$\|\alpha e\| = |\alpha| \, \|e\| \qquad (3.2)$$
$$\|e_1 + e_2\| \leqslant \|e_1\| + \|e_2\|$$

where equality in the first of Eqs. (3.2) may occur only with $e = 0$, and where α is a number. The L_∞ error measure of Eq. (3.1) is termed *local* or *pointwise*. Observe in Fig. 3.1c that a good local approximation of y does not foretell the quality of the approximation of y'.

Another way to measure the error between y and $\tilde{y}$ in, say, the interval $0 \leqslant x \leqslant 1$, is to do it *globally* by the L_2 error measure

$$\|y - \tilde{y}\|_0 = \left[\int_0^1 (y - \tilde{y})^2 \, dx \right]^{1/2} \qquad (3.3)$$

that is appropriate for all *square integrable* functions. A zero appears in the subscript of $\|\cdot\|_0$ in Eq. (3.3) to point out that $e(x) = y - \tilde{y}$ itself is square integrated and that none of its higher-order derivatives are included in the norm. We accept $\|\cdot\|_0$ as an error measure by the fact, readily verified, that it satisfies the three basic properties of the norm in Eq. (3.2).

In many instances y' is of the same physical interest as y and its accuracy is as important. To include y' in the error measure, we introduce the $\|\cdot\|_1$ error norm

$$\|y - \tilde{y}\|_1 = \left\{ \int_0^1 [(y' - \tilde{y}')^2 + (y - \tilde{y})^2] \, dx \right\}^{1/2} \qquad (3.4)$$

or even more generally

$$\|y - \tilde{y}\|_1 = \left\{ \int_0^1 [p(x)(y' - \tilde{y}')^2 + q(x)(y - \tilde{y})^2] \, dx \right\}^{1/2} \qquad (3.5)$$

in which $p(x) > 0$, $p(x) = 0$ only at distinct points, and $q(x) \geqslant 0$. The inclusion of e' in the integral is indicated by the subscript 1 of $\|\cdot\|_1$. This norm is destined to play a central role in the rest of our discussion and has a meaningful physical implication:

$$\|y\|_1{}^2 = \int_0^1 [p(x)y'^2 + q(x)y^2]\, dx \tag{3.6}$$

is twice the energy stored in an elastically supported string under the tension $p(x)$ as a result of its displacement $y(x)$. Accordingly, $\|e\|_1{}^2$ is the energy supposedly stored in the string by the fictitious displacement $e(x) = y(x) - \tilde{y}(x)$, and $\|e\|_1{}^2$ is fittingly termed the energy error, and $\|\cdot\|_1$ the energy norm. When $p(x) > 0$ and $q(x) > 0$, $\|e\|_1$ in Eq. (3.5) is a true norm in the sense of Eqs. (3.2). To accommodate physics, we also permit $q(x) = 0$ in Eq. (3.5) or (3.6). But when $q(x) = 0$, $\|e\|_1$ vanishes not only for $e = 0$ but also when e is any constant since then $e' = 0$, and the first of the requirements (3.2) put on the norm is not fulfilled. If we restrict ourselves, however, to *continuous*, C^0, y and $\tilde{y}$ and assume that *at least at one point* $e = y - \tilde{y} = 0$, then the continuity assumption guarantees that if $y - \tilde{y} = \text{const.}$, then this constant is necessarily zero and $\|y - \tilde{y}\|_1 = 0$ only when $y - \tilde{y} = 0$, even when $q(x) = 0$.

Discontinuities are allowed in $p(x)$, and $q(x)$ may even be a delta function as when the string is elastically supported at distinct points by isolated springs.

Functions whose magnitude we wish to measure energetically must, of course, be of finite energy. Note carefully that continuity is essential in such functions but not sufficient. Take for example $y = \sqrt{x}$, $0 \leqslant x \leqslant 1$, which is certainly continuous in this interval but is not of a finite $\|\cdot\|_1$ energy norm

$$\|y - \tilde{y}\|_1{}^2 = \int_0^1 (y' - \tilde{y}')^2\, dx \tag{3.7}$$

In all the following physical applications we shall assume that y, whatever its physical meaning is, is of finite energy. As for the *trial* or *comparison* function $\tilde{y}$ whose closeness to y we decide to measure energetically, very often it is chosen in the form of a *piecewise polynomial*. In this case satisfaction of the proper continuity requirements on $\tilde{y}$ to make $\|\tilde{y}\|_1$ finite is of central importance.

A discontinuity in $\tilde{y}$ can be physically construed in different ways. If $\tilde{y}$ is taken, as in Fig. 3.2a, to consist of horizontally cut string pieces (i.e., $\tilde{y}$ is single valued at the points of discontinuity) and the integral in Eq. (3.7) is summed over the string segments, then the energy of $\tilde{y}$ amounts to zero

Fig. 3.2 Two kinds of energetically inadmissible discontinuities in a string.

(a)   (b)

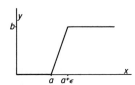

Fig. 3.3   As $\varepsilon \to 0$ the elastic energy of the string goes to ∞.

and for any such $\tilde{y}$

$$\|y - \tilde{y}\|_1{}^2 = \int_0^1 y'^2 \, dx \tag{3.8}$$

which is, needless to say, useless. If the discontinuity of $\tilde{y}$ is of the type shown in Fig. 3.2b, the string being physically continuous but stretched stepwise (i.e., $\tilde{y}$ is multivalued at the points of discontinuity) then $\|\tilde{y}\|_1$ is infinite, which is also inadmissible. Indeed, for the string in Fig. 3.3

$$\|\tilde{y}\|_1{}^2 = \int_a^{a+\varepsilon} \frac{b^2}{\varepsilon^2} \, dx = \frac{b^2}{\varepsilon} \tag{3.9}$$

and $\|\tilde{y}\|_1 \to \infty$ as $\varepsilon \to 0$. Thus, all candidates $\tilde{y}$ to approximate y, with $e(x) = y(x) - \tilde{y}(x)$ measured by the energy norm $\|\cdot\|_1$ in Eq. (3.7), must satisfy the conditions

(i) $\tilde{y}$ is continuous and of finite energy;
(ii) $y - \tilde{y} = 0$ at least at one point.

The energy norm $\|\cdot\|_2$ of the elastically supported thin elastic beam is

$$\|y\|_2 = \left\{ \int_0^1 [p(x)y''^2 + q(x)y^2] \, dx \right\}^{1/2} \tag{3.10}$$

in which $p(x) > 0$ and $q(x) \geq 0$, and where the subscript 2 points to the appearance of y'' in the integral. Both $p(x)$ and $q(x)$ in Eq. (3.10) may change suddenly, with $q(x)$ possibly even only a point support, as in Fig. 3.4. In this case, and with a load that is allowed to act at a point, the deflection y of the beam is C^1; it is continuous and has a continuous first derivative, and $\|y\|_2$ is finite. When the elastic support is absent and the beam of constant (say unit) properties, the energy error in $y - \tilde{y}$ becomes

$$\|y - \tilde{y}\|_2{}^2 = \int_0^1 (y'' - \tilde{y}'')^2 \, dx \tag{3.11}$$

The appearance of $\tilde{y}''^2$ in $\|e\|_2$ restricts the choice of $\tilde{y}$ to those functions that are C^1. A discontinuity in $\tilde{y}'$ can be physically interpreted as either that in Fig. 3.5a or b. In Fig. 3.5a, $\tilde{y}$ consists of straight bars that are pin jointed,

Fig. 3.4   Simply supported beam, having an abrupt change of thickness, elastically supported at a point.

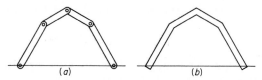

Fig. 3.5   Two kinds of energetically inadmissible discontinuities in a beam.

while that in Fig. 3.5b consists of straight beam segments that abruptly change direction. No one of these approximations is energetically admissible. In case a, $\|y - \tilde{y}\|_2 = \|y\|_2$, while in case b, $\|\tilde{y}\|_2 = \infty$.

To use the error measure (3.11) both y and $\tilde{y}$ have to be finite in the $\|\cdot\|_2$ norm. In particular, they have to be continuous and have a continuous first derivative. But this is not enough to render $\|y - \tilde{y}\|_2$ an acceptable error measure. Still $\|y - \tilde{y}\|_2 = 0$ happens for all $y - \tilde{y} = a_0 + a_1 x$. This is avoided by the additional restriction on y that $y = \tilde{y}$ at least at *two* points, or $y = \tilde{y}$ and $y' = \tilde{y}'$ at one point. For $\tilde{y}$ to be a candidate for approximating y in the energy norm $\|\cdot\|_2$, it must satisfy the conditions that

(i) $\tilde{y}$ and $\tilde{y}'$ be continuous and of finite energy;
(ii) $y = \tilde{y}$ at least at two points or $y = \tilde{y}$ and $y' = \tilde{y}'$ at one point.

2.   Principle of Minimum Potential Energy

Reconsider the second-order two-point boundary value problem

$$-(p(x)y')' + q(x)y = f(x), \qquad 0 < x < 1$$
$$y(0) = 0, \qquad y'(1) = 0 \tag{3.12}$$

with the companion energy norm

$$\|y\|_1{}^2 = \int_0^1 [p(x)y'^2 + q(x)y^2]\,dx \tag{3.13}$$

Let $\tilde{y}$ be an energetically admissible comparison function and $y(0) = \tilde{y}(0) = 0$. We choose to measure the error $y - \tilde{y}$ between the trial function $\tilde{y}$ and the solution y to the boundary value problem in the $\|\cdot\|_1$ norm, and have that

$$\|y - \tilde{y}\|_1{}^2 = \int_0^1 [p(x)(y' - \tilde{y}')^2 + q(x)(y - \tilde{y})^2]\,dx \tag{3.14}$$

Theoretically, we are now able to distinguish between good and better $\tilde{y}$ according to the value of $\|y - \tilde{y}\|_1$ in Eq. (3.14), and possibly institute a procedure for the systematic search of a sequence of approximations $\tilde{y}_0$,

$\tilde{y}_1, \ldots, \tilde{y}_N$ that *converge in the energy* to the exact solution y:

$$\lim_{N \to \infty} \|y - \tilde{y}_N\|_1 = 0 \qquad (3.15)$$

However, at this point this is only wishful thinking because the unknown function y is still included in Eq. (3.14). What we plan to do in this section is to separate $\|y - \tilde{y}\|_1^2$ into two parts; one including only y and the other only $\tilde{y}$. This is possible only with a special choice of energy norm and in this way, this norm imposes itself upon us instead of, say, the $\|\cdot\|_0$ or $\|\cdot\|_\infty$ norms. Subsequently, after energy convergence has been determined, we shall also discuss the convergence, or the effectiveness, of the $\tilde{y}$ approximation in other norms.

The integrand in Eq. (3.14) can be rearranged in the form

$$\|y - \tilde{y}\|_1^2 = \int_0^1 \{p(x)[\tilde{y}'^2 - y'^2 + 2y'(y' - \tilde{y}')] + q(x)[y^2 - 2y\tilde{y} + \tilde{y}^2]\} \, dx \qquad (3.16)$$

Now, if in addition to our basic assumption that both y and $\tilde{y}$ are continuous we also assume that py' is continuous—an assumption of valid physical significance—then the term $py'(y' - \tilde{y}')$ in Eq. (3.16) may be integrated by parts in the form

$$\int_0^1 py'(y' - \tilde{y}') \, dx = -\int_0^1 (py')'(y - \tilde{y}) \, dx + py'(y - \tilde{y}) \Big|_0^1 \qquad (3.17)$$

At the end point $x = 0$, we have that $y - \tilde{y} = 0$, while at the point $x = 1$, we have that $y' = 0$, and hence the boundary term in Eq. (3.17) disappears. Substitution of Eq. (3.17) back into Eq. (3.16) for $\|y - \tilde{y}\|_1^2$, together with some rearrangements, results in

$$\|y - \tilde{y}\|_1^2 = \int_0^1 \{p(x)(\tilde{y}'^2 - y'^2) + q(x)(\tilde{y}^2 - y^2)$$
$$+ 2(y - \tilde{y})[-(p(x)y')' + q(x)y]\} \, dx \qquad (3.18)$$

which, since $-(py')' + qy = f$, turns into

$$\|y - \tilde{y}\|_1^2 = \int_0^1 [p(x)(\tilde{y}'^2 - y'^2) + q(x)(\tilde{y}^2 - y^2) + 2(y - \tilde{y})f(x)] \, dx \qquad (3.19)$$

Equation (3.19) prompts us to introduce the *total potential energy*

$$\pi(y) = \tfrac{1}{2} \int_0^1 [p(x)y'^2 + q(x)y^2 - 2f(x)y] \, dx \qquad (3.20)$$

associated with the two-point boundary value problem, with which $\|y - \tilde{y}\|_1^2$ in Eq. (3.19) can be written as

$$\tfrac{1}{2}\|y - \tilde{y}\|_1^2 = \pi(\tilde{y}) - \pi(y) \qquad (3.21)$$

and we achieve our goal of separating $\|y - \tilde{y}\|_1^2$ into an expression $\pi(y)$, including only y and an expression $\pi(\tilde{y})$, involving only $\tilde{y}$.

The total potential energy $\pi(y)$ in Eq. (3.20) is a *functional*. For any admissible *function* y going into it, it delivers a *number* $\pi(y)$. Since $\|y - \tilde{y}\|_1 \geqslant 0$, we have from Eq. (3.21) that

$$\pi(\tilde{y}) \geqslant \pi(y) \qquad (3.22)$$

where equality holds only for $y = \tilde{y}$; meaning that any admissible trial function $\tilde{y}$ has a total potential energy higher than that corresponding to y. The true solution y to the boundary value problem in Eq. (3.12) *minimizes* the total potential energy. This is the *principle of minimum potential energy*.

By virtue of Eq. (3.21), the minimization of $\pi(\tilde{y})$ equals the minimization of $\|y - \tilde{y}\|_1$ and the two-point boundary value problem can be formulated *variationally* as the requirement to minimize $\pi(\tilde{y})$ over the class C^0 of functions that have square integrable first derivatives and that satisfy the conditions $y(0) = \tilde{y}(0) = 0$. It is interesting that the class of comparison functions $\tilde{y}$, over which the minimum of $\pi(\tilde{y})$ is sought, need satisfy only *one* boundary condition, namely, $y(0) = 0$, and not the other $y'(1) = 0$. The latter boundary condition is included in the minimization process in the same way the differential equation is included in it and is therefore termed a *natural* boundary condition in contrast to $y(0) = 0$, which is an *essential* boundary condition. Among physicists and engineers these conditions are also known as *dynamical* (natural) and *geometrical* (essential) because of the physical meaning y and y' carry in the mechanics of solids.

The fact that only the lower-order boundary derivatives of y need be exactly satisfied in the variational formulation of the boundary value problem is of significant importance in the numerical approximation methods based on the variational formulation.

The total potential energy in Eq. (3.20) was derived under the assumption that py' is continuous. It can happen, as when the string is point loaded, that there is a sudden jump in py' with y still of finite energy and physically plausible. The total potential energy $\pi(y)$ in Eq. (3.20) also covers this case except that $f(x)$ is a point source or load. For a formal demonstration of this consider

$$\begin{aligned} -y'' &= P(x_0), \qquad 0 < x < 1 \\ y(0) &= y(1) = 0 \end{aligned} \qquad (3.23)$$

where $P(x_0)$ is a point force of strength P at $x = x_0$, or

$$y'(x_0^+) - y'(x_0^-) = -P \qquad (3.24)$$

We associate with the boundary value problem (3.23) the energy error norm $\|\cdot\|_1$:

$$\|y - \tilde{y}\|_1^2 = \int_0^1 (y' - \tilde{y}')^2\, dx = \int_0^1 [\tilde{y}'^2 - y'^2 + 2y'(y' - \tilde{y}')]\, dx \quad (3.25)$$

as in Eqs. (3.13) and (3.16). Integration by parts of $y'(y' - \tilde{y}')$ now has to take into account the discontinuity of y', and

$$\int_0^1 y'(y' - \tilde{y}')\, dx = -\int_0^{x_0^-} y''(y - \tilde{y})\, dx - \int_{x_0^+}^1 y''(y - \tilde{y})\, dx$$

$$+ y'(y - \tilde{y})\Big|_0^{x_0^-} + y'(y - \tilde{y})\Big|_{x_0^+}^1 \quad (3.26)$$

Because $y'' = 0$ when $x \neq x_0$, we are left with

$$\int_0^1 y'(y' - \tilde{y}') = [y'(x_0^-) - y'(x_0^+)][y(x_0) - \tilde{y}(x_0)]$$

$$= P[y(x_0) - \tilde{y}(x_0)] \quad (3.27)$$

and the total potential energy becomes

$$\pi(y) = \tfrac{1}{2} \int_0^1 y'^2\, dx - Py(x_0) \quad (3.28)$$

as in Eq. (3.20) when $f(x) = P(x_0)$.

3.   More General Boundary Conditions

The energy norm and the corresponding total potential energy associated with a boundary value problem depend on both the differential equation and the boundary conditions. To develop the potential energy for the more general boundary conditions that may accompany the differential equation (3.12), we assume it is describing the stationary temperature distribution in a rod radiating into a space of ambient temperatures y_a at $x = 0$ and y_b at $x = 1$, with corresponding (positive) radiation coefficients h_0 and h_1. Accordingly, at points $x = 0$ and $x = 1$ we have the boundary conditions

$$\begin{aligned} py' &= h_0(y - y_a) \qquad \text{at} \quad x = 0 \\ -py' &= h_1(y - y_b) \qquad \text{at} \quad x = 1 \end{aligned} \quad (3.29)$$

or

$$\begin{aligned} -py' + h_0 y &= h_0 y_a = c_0 \\ py' + h_1 y &= h_1 y_b = c_1 \end{aligned} \quad (3.30)$$

When $h = \infty$, Eqs. (3.30) reduce to $y = y_a$, and $y = y_b$, the conditions of prescribed temperature, while $h = 0$ reduces Eqs. (3.30) to $y' = 0$; the condition of insulated ends. With the more general boundary conditions in Eq. (3.30), a boundary energy term need be added to the energy norm in order to permit the derivation of a principle of minimum potential energy, and we write

$$\|y\|_1{}^2 = \int_0^1 [p(x)y'^2 + q(x)y^2]\,dx + h_1 y^2(1) + h_0 y^2(0) \qquad (3.31)$$

The energy error in any trial function $\tilde{y}$ with square integrable derivatives becomes with Eq. (3.31)

$$\|y - \tilde{y}\|_1{}^2 = \int_0^1 [p(x)(y' - \tilde{y}')^2 + q(x)(y - \tilde{y})^2]\,dx$$
$$+ h_1[y(1) - \tilde{y}(1)]^2 + h_0[y(0) - \tilde{y}(0)]^2 \qquad (3.32)$$

In case either $h_1 \neq 0$ or $h_2 \neq 0$, $\|y - \tilde{y}\|_1$ is a valid norm even if $q(x) = 0$, by virtue of the boundary terms, and we need not impose any additional conditions on $\tilde{y}$; both boundary conditions in Eq. (3.30) are natural.

To get the total potential energy expression from Eq. (3.32), we proceed as in Section 2 except that the present boundary term need be handled with more care. Integration by parts similar to that in Eq. (3.17) yields here

$$p(x)y'(y - \tilde{y})|_0^1 = p_1 y_1{}'(y_1 - \tilde{y}_1) - p_0 y_0{}'(y_0 - \tilde{y}_0) \qquad (3.33)$$

where p_0 and y_0 stand for the values of p and y at $x = 0$. Since $p_1 y_1{}' = c_1 - h_1 y_1$ and $-p_0 y_0{}' = c_0 - h_0 y_0$, Eq. (3.33) becomes

$$p(x)y'(y - \tilde{y})|_0^1 = (c_1 - h_1 y_1)(y_1 - \tilde{y}_1) + (c_0 - h_0 y_0)(y_0 - \tilde{y}_0)$$
$$= c_1(y_1 - \tilde{y}_1) - h_1 y_1{}^2 + h_1 y_1 \tilde{y}_1 + c_0(y_0 - \tilde{y}_0) \qquad (3.34)$$
$$- h_0 y_0{}^2 + h_0 y_0 \tilde{y}_0$$

This, combined with the boundary terms in Eq. (3.32), leads to the total potential energy

$$\pi(y) = \tfrac{1}{2} \int_0^1 [p(x)y'^2 + q(x)y^2 - 2f(x)y]\,dx$$
$$- c_1 y(1) - c_0 y(0) + \tfrac{1}{2}h_1 y^2(1) + \tfrac{1}{2}h_0 y^2(0) \qquad (3.35)$$

and Eq. (3.32) is compactly written as $\tfrac{1}{2}\|y - \tilde{y}\|_1{}^2 = \pi(\tilde{y}) - \pi(y)$. The variational formulation of the stationary temperature distribution problem in the rod, given by the differential equation in Eq. (3.12) and the boundary conditions in Eq. (3.30), becomes that of requiring the minimization of $\pi(\tilde{y})$ in Eq. (3.35) over the class of all continuous functions with finite energy but *with $\tilde{y}$ not required to satisfy any boundary conditions.*

In case $h_0 = \infty$, the boundary condition at $x = 0$ reduces to $y(0) = y_a$, and the trial function $\tilde{y}$ must satisfy the condition $y(0) = \tilde{y}(0) = y_a$ in order to permit the derivation of a potential energy from the energy norm. In case both $h_0 = \infty$ and $h_1 = \infty$, the boundary term in Eq. (3.32) drops out and $\tilde{y}$ is made to satisfy $y(0) = \tilde{y}(0) = y_a$ and $y(1) = \tilde{y}(1) = y_b$. The approximate satisfaction of these conditions can be achieved with a very large h but is computationally risky.

4. Complementary Variational Principles

In Section 2 we associated with the two-point boundary value problem the energy norm $\|y\|_1^2$ given in Eq. (3.13) and, consequently, a total potential energy $\pi(y)$ given in Eq. (3.20), and got that for any continuous trial function $\tilde{y}$ of finite energy, satisfying the essential boundary condition $\tilde{y}(0) = 0$, $\pi(\tilde{y}) \geqslant \pi(y)$, where y is the true solution to the two-point boundary value problem in Eq. (3.12). Thus the value of $\pi(\tilde{y})$ constitutes an *upper* bound on $\pi(y)$. In this section we intend to introduce another, *complementary*, functional that will provide a *lower* bound on $\pi(y)$. To this end we introduce the complementary trial function $\bar{y}$ that is required to satisfy the differential equation and the boundary condition at the end $x = 1$:

$$-(p(x)\bar{y}')' + q(x)\bar{y} = f(x), \qquad 0 < x < 1, \qquad \bar{y}'(1) = 0 \qquad (3.36)$$

and $\bar{y}$ is seen to be required to satisfy the conditions $\tilde{y}$ is exempt from but is not required to fulfill the condition $\bar{y}(0) = 0$ that $\tilde{y}$ is required to satisfy in the principle of minimum potential energy. We obviously have that $\|y - \bar{y}\|_1^2 \geqslant 0$ and proceed to separate $\|y - \bar{y}\|_1^2$ into expressions involving only y and only $\bar{y}$. Upon the expansion of the integrand of $\|y - \bar{y}\|_1^2$, we get the same expression as in Eq. (3.16) except that $\tilde{y}$ is replaced by $\bar{y}$. We also integrate by parts the term $p(x)y'(y' - \bar{y}')$ in a different manner. Here

$$\int_0^1 p(x)y'(y' - \bar{y}')\,dx = -\int_0^1 y[p(x)(y' - \bar{y}')]'\,dx + p(x)(y' - \bar{y}')y\Big|_0^1 \qquad (3.37)$$

and because $y(0) = 0$, and $y'(1) = \bar{y}'(1) = 0$, the boundary term in Eq. (3.37) vanishes. Moreover, since $\bar{y}$ satisfies the differential equation (3.36), we have that

$$-(py')' + (p\bar{y}')' = -q(y - \bar{y}) \qquad (3.38)$$

and consequently,

$$\|y - \bar{y}\|_1^2 = \int_0^1 [p(x)(\bar{y}'^2 - y'^2) + q(x)(\bar{y}^2 - y^2)]\,dx \qquad (3.39)$$

With the *complementary potential energy* $\psi(y)$ defined as

$$\psi(y) = -\tfrac{1}{2} \int_0^1 \left[p(x)y'^2 + q(x)y^2 \right] dx \tag{3.40}$$

Equation (3.39) can be concisely written as

$$\tfrac{1}{2} \| y - \bar{y} \|_1^2 = \psi(y) - \psi(\bar{y}) \geqslant 0 \tag{3.41}$$

where equality holds only when $y = \bar{y}$.

It can be readily verified that for the solution y to the boundary value problem in Eq. (3.12)

$$\pi(y) = \psi(y) = -\tfrac{1}{2} \int_0^1 \left[p(x)y'^2 + q(x)y^2 \right] dx \tag{3.42}$$

and Eq. (3.41) implies therefore that

$$\pi(y) \geqslant \psi(\bar{y}) \tag{3.43}$$

meaning that $\psi(\bar{y})$ is lower than $\pi(y)$ for any admissible $\bar{y}$ (i.e., satisfying Eqs. (3.36)]. Equation (3.43) embodies the principle of *maximum complementary energy*; the true solution y maximizes $\psi(\bar{y})$ and inasmuch as

$$\pi(\tilde{y}) \geqslant \pi(y) \geqslant \psi(\bar{y}) \tag{3.44}$$

we have from Eq. (3.21) that

$$\tfrac{1}{2} \| y - \tilde{y} \|_1^2 \leqslant \pi(\tilde{y}) - \psi(\bar{y}) \tag{3.45}$$

and a computable bound is established on $\| y - \tilde{y} \|_1$. The practical realization of this bound is not always feasible because of the stringent conditions on $\bar{y}$. A systematic search for $\tilde{y}$ trial functions to minimize $\| y - \tilde{y} \|_1$ is the object of the method of finite elements to be expounded in detail in the next chapter.

To numerically substantiate the ideas of this section, we choose to apply them to the two-point boundary value problem $-y'' + y = 1$, $0 < x < 1$, with $y(0) = 0$, $y'(1) = 0$, that admits the solution $y = c_1 e^x + c_2 e^{-x} + 1$, where $c_1 = -1/(1 + e^2)$ and $c_2 = -e^2/(1 + e^2)$, shown in Fig. 3.6. By direct calculations, we find that for this y, $\pi(y) = c_1 = -0.1192$. As an admissible trial function for the principle of minimum potential energy and the principle of complementary energy, we choose $\tilde{y} = x$, $\bar{y} = 1$, respectively. Notice that $\tilde{y}$ satisfies the boundary condition $\tilde{y}(0) = 0$ and that $\bar{y}$ satisfies both the differential equation and boundary condition $\bar{y}'(1) = 0$. The approximation $\bar{y} = 1$ is admittedly useless but a better one does not readily present itself. With $\tilde{y} = x$ we get $\pi(\tilde{y}) = \tfrac{1}{6}$, while from $\bar{y} = 1$ we get $\psi(\bar{y}) = -0.5$. Also $\| y - \tilde{y} \|_1^2 = 0.5717$ while from Eq. (3.45) we get that $\| y - \tilde{y} \|_1^2 \leqslant 1.333$. The approximation $y = x$ can be improved, and this hints of things to come, by taking $\tilde{y} = \alpha x$ and minimizing $\pi(\tilde{y})$ with respect to α in order to minimize

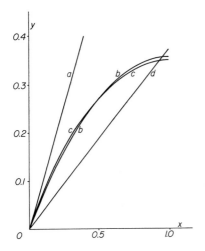

Fig. 3.6 Trial functions: (a) $\bar{y} = x$, (b) $\bar{y} = \frac{5}{14}(2x - x^2)$, (d) $\bar{y} = \frac{3}{8}x$, and (c) exact solution of $-y'' + y = 1$, $0 < x < 1$, $y(0) = y'(1) = 0$.

$\|y - \bar{y}\|_1$. Presently $\pi(\bar{y}) = \frac{1}{2}\alpha(\frac{4}{3}\alpha - 1)$, the best α is given by $d\pi/d\alpha = \frac{8}{3}\alpha - 1 = 0$, and $\alpha = \frac{3}{8}$. That $\bar{y} = \frac{3}{8}x$ is a better approximation than $\bar{y} = x$ can be clearly seen in Fig. 3.6. Indeed with this α, $\pi(\bar{y}) = -\frac{3}{32} = -0.09375$, lower than the $\pi(\bar{y}) = 0.1667$ gotten with $\bar{y} = x$, but still above $\pi(y) = -0.1192$. In case the shape of the solution can be deduced from, say, previous experience or any other analytic consideration, then a still better trial function can be guessed. In our case, the curve for y in Fig. 3.6 suggests the trial function $\bar{y} = \alpha x(2 - x)$ and the best α is found from the minimization of $\pi(\bar{y}) = \frac{14}{15}\alpha^2 - \frac{2}{3}\alpha$, yielding $\alpha = \frac{5}{14}$ and $\bar{y} = \frac{5}{14}x(2 - x)$, which is truly an excellent approximation according to Fig. 3.6. For this last trial function, $\pi(\bar{y}) = \frac{15}{42} = -0.11905$, which is, notice, still slightly higher than $\pi(y) = -0.1192$.

The total potential energy and the energy error are usually not of direct interest to the engineer or scientist solving the problem. As far as the approximation errors are concerned we are more interested in $\|y - \bar{y}\|_\infty$ or even $\|y' - \bar{y}'\|_\infty$. How to relate these to the energy error is delayed to the discussion on finite element approximations.

5. Euler–Lagrange Equations

The thrust of Sections 2 and 3 was to prove that the solution to the boundary value problem minimizes the total potential energy. Next we approach the principle of total potential energy from the other end; using the techniques of the calculus of variations, we show that the function that minimizes

$\pi(\tilde{y})$ in Eq. (3.20) is actually the solution of the boundary value problem (3.12). Let y be the function that minimizes $\pi(y)$ in Eq. (3.20). We perturb y by a *virtual displacement* $\varepsilon\tilde{y}$ in which ε is a scalar variable and $\tilde{y}$ a continuous function of finite energy and such that $\tilde{y}(0) = 0$. The total potential energy at the disturbed state $y + \varepsilon\tilde{y}$ is

$$\pi(y + \varepsilon\tilde{y}) = \tfrac{1}{2}\int_0^1 \left[p(x)(y' + \varepsilon\tilde{y}')^2 + q(x)(y + \varepsilon\tilde{y})^2 - 2f(y + \varepsilon\tilde{y})\right]dx \quad (3.46)$$

Upon expansion it becomes

$$\pi + \delta\pi + \delta^2\pi = \tfrac{1}{2}\int_0^1 \left[p(x)y'^2 + q(x)y^2 - 2f(x)y\right]dx$$
$$+ \varepsilon\int_0^1 \left[p(x)y'\tilde{y}' + q(x)y\tilde{y} - f(x)\tilde{y}\right]dx$$
$$+ \tfrac{1}{2}\varepsilon^2\int_0^1 \left[p(x)\tilde{y}'^2 + q(x)\tilde{y}^2\right]dx \quad (3.47)$$

The integral before ε is the *first variation* $\delta\pi$ of π, while that before ε^2 is the *second variation* $\delta^2\pi$ of π and is equal to $\tfrac{1}{2}\|\tilde{y}\|_1^2$. Notice the analogy between differentiation and first variation, that for y'^2 yields $2y'\tilde{y}'$, for y^2 yields $2y\tilde{y}$, and for y yields $\tilde{y}$.

Whatever the choice of $\tilde{y}$ in $y + \varepsilon\tilde{y}$, the total potential energy is minimal at $\varepsilon = 0$. A necessary condition for this minimum is that $d\pi/d\varepsilon = 0$ at $\varepsilon = 0$, leading to

$$\delta\pi = \int_0^1 \left[p(x)y'\tilde{y}' + q(x)y\tilde{y} - f(x)\tilde{y}\right]dx = 0 \quad (3.48)$$

or the condition that the first variation of π vanish for *any* admissible trial function $\tilde{y}$. Integration by parts transforms Eq. (3.48) into

$$\int_0^1 \left[-(p(x)y')' + q(x)y - f(x)\right]\tilde{y}\,dx + p(1)y'(1)\tilde{y}(1) = 0 \quad (3.49)$$

since $\tilde{y}(0) = 0$. It is a fundamental theorem of the calculus of variations that if the integral, plus boundary term, vanishes for *any* admissible $\tilde{y}$, then this can occur only if

$$-(p(x)y')' + q(x)y = f(x), \qquad y'(1) = 0 \quad (3.50)$$

which are the *Euler–Lagrange equations* associated with $\pi(y)$: the differential equation and natural boundary condition we started with. That y satisfies the condition $y(0) = 0$ we knew beforehand since the class of functions from which the candidates for minimizing $\pi(\tilde{y})$ are drawn included this condition. Thus, the function y that minimizes $\pi(\tilde{y})$ is the solution to our two point boundary value problem.

The fundamental theorem of the calculus of variations, claiming that if

$$\int_0^1 z(x)\tilde{y}(x)\,dx = 0 \quad (3.51)$$

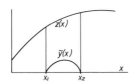

Fig. 3.7   Choice of $\bar{y}(x)$ to make $z(x)y(x)$ positive.

where $z(x) = -(p(x)y')' + q(x)y - f(x)$ holds for any energetically admissible $\bar{y}$, then $z(x) = 0$, is proved as follows: Assume $z(x)$ continuous and of the same sign, say positive, at the interval $x_0 \leqslant x \leqslant x_1$, as in Fig. 3.7. A choice of $\bar{y}$ can be made for it, as in Fig. 3.7, rendering the integral in Eq. (3.49) positive, in contradiction to Eq. (3.51).

6.   Total Potential Energy of the Thin Elastic Beam

The differential equation governing the deflection y of a thin elastic beam which is elastically supported is given in Eq. (2.29). Added to this equation are the common homogeneous boundary conditions

$$
\begin{array}{lll}
y = 0, & y'' = 0 & \text{at a simply supported end} \\
y = 0, & y' = 0 & \text{at a clamped end} \\
y' = 0, & (py'')' = 0 & \text{at a point of symmetry} \\
y'' = 0, & (py'')' = 0 & \text{at a free end}
\end{array}
\qquad (3.52)
$$

We select a comparison (trial) function $\bar{y}$ for the beam that is C^1 (i.e., continuous and with a continuous first derivative) and that satisfies the conditions $\bar{y} = 0$, $\bar{y}' = 0$ wherever $y = y' = 0$ occurs. These are the essential boundary conditions of the beam problem. The remaining higher-order conditions are natural.

From the definition of $\|\cdot\|_2$ in Eq. (3.10) we have that

$$
\|y - \bar{y}\|_2^2 = \int_0^1 \{p(x)[-y''^2 + \bar{y}''^2 + 2y''(y'' - \bar{y}'')] \\
+ q(x)[y^2 - 2y\bar{y} + \bar{y}^2]\} \, dx
\qquad (3.53)
$$

Two successive integrations by parts of $p(x)y''(y'' - \bar{y}'')$ yields

$$
\int_0^1 p(x)y''(y'' - \bar{y}'') \, dx = \int_0^1 [p(x)y'']''(y - \bar{y}) \, dx - [p(x)y'']'(y - \bar{y})\Big|_0^1 \\
+ p(x)y''(y' - \bar{y}')\Big|_0^1
\qquad (3.54)
$$

To see that the boundary terms in Eq. (3.54) vanish when y satisfies one of the boundary conditions in Eq. (3.52), and $\tilde{y}$ and $\tilde{y}'$ are equal to y and y' wherever these are prescribed, we observe in Eq. (3.52) that wherever $(p(x)y'')'$ is not prescribed, $y = \tilde{y} = 0$, and wherever y'' is not prescribed, $y' = \tilde{y}' = 0$. On the other hand, at points where y is not prescribed, $(py'')' = 0$, and at points where y' is not prescribed, $y'' = 0$.

Reintroduction of Eq. (3.54) into Eq. (3.53), while remembering that $(py'')'' + qy = f$, produces

$$\tfrac{1}{2}\|y - \tilde{y}\|_2^2 = \pi(\tilde{y}) - \pi(y) \geqslant 0 \tag{3.55}$$

where

$$\pi(y) = \tfrac{1}{2}\int_0^1 \left[p(x)y''^2 + q(x)y^2 - 2f(x)y \right] dx \tag{3.56}$$

is the total potential energy of the elastic beam.

7. Indefinite Variational Principles

The calculus of variations technique of Section 5 enables us to construct and verify functionals whose Euler–Lagrange equations are the differential equations we are seeking to solve even in case these functionals do not possess a true minimum, but merely a more general extremum point. A noteworthy example for this presents itself when the beam equation is split up into the two Eqs. (2.38). The boundary conditions in Eq. (3.52) become, in term of y and M

$$
\begin{aligned}
y &= 0, & M &= 0 \\
y &= 0, & y' &= 0 \\
y' &= 0, & M' &= 0 \\
M &= 0, & M' &= 0
\end{aligned}
\tag{3.57}
$$

We associate with Eqs. (2.38) and the boundary conditions (3.57) the *two field* functional

$$\pi(y, M) = \int_0^1 \left[\frac{1}{2p(x)}M^2 - \frac{1}{2}q(x)y^2 - y'M' + fy \right] dx \tag{3.58}$$

where $p(x) > 0$. Forming $\delta\pi = 0$ with respect to M and y, we get

$$\int_0^1 \left[\frac{1}{p(x)}M + y'' \right] \tilde{M}\, dx - y'\tilde{M}\Big|_0^1 = 0 \tag{3.59}$$

and

$$\int_0^1 [-qy + M'' + f]\tilde{y}\,dx - \tilde{y}M' \Big|_0^1 = 0 \qquad (3.60)$$

respectively, where $\tilde{y}$ and $\tilde{M}$ are continuous trial functions for y and M. By the argument of Section 5, the expression in brackets must vanish yielding the two differential equations $M + p(x)y'' = 0$ and $M'' - q(x)y + f(x) = 0$. From the boundary terms in Eqs. (3.59) and (3.60), we conclude that where M is not prescribed and $\tilde{M}$ is free to vary, $y' = 0$ is a natural boundary condition, while at points where y is not prescribed and $\tilde{y}$ is allowed to vary, $M' = 0$ is a natural condition. The essential boundary conditions that the candidate pair $(\tilde{M}, \tilde{y})$ for extremizing $\pi(\tilde{y}, \tilde{M})$ need satisfy are $\tilde{y} = 0$, $\tilde{M} = 0$ whenever $y = 0$ and $M = 0$. The conditions $y' = 0$ and $M' = 0$ are natural.

Fig. 3.8 Clamped free beam point loaded at the tip.

As an example of the computational use of indefinite variational principles, consider the unit cantilever beam (Fig. 3.8) point loaded by a unit force at its tip. The solution to this problem is $y = \frac{1}{2}x^2 - \frac{1}{6}x^3$, and $M = x - 1$. We propose to seek an approximate pair $y = \alpha x$ and $M = \beta(1 - x^2)$ for the deflection and moment distribution along the beam through the extremization of $\pi(\tilde{y}, \tilde{M})$ in Eq. (3.58). With this choice of y and M, and since here $p(x) = 1$, $q(x) = 0$,

$$\pi(\tilde{y}, \tilde{M}) = \int_0^1 \left[\tfrac{1}{2}\beta^2(x^2 - 1)^2 + 2\alpha\beta x\right] dx + \alpha \qquad (3.61)$$

where the last α comes from the point load. Integration yields

$$\pi(\tilde{y}, \tilde{M}) = \tfrac{4}{15}\beta^2 + \alpha\beta + \alpha \qquad (3.62)$$

and

$$\frac{\partial \pi}{\partial \alpha} = +\beta + 1 = 0, \qquad \frac{\partial \pi}{\partial \beta} = \frac{8}{15}\beta + \alpha = 0$$

from which $\beta = -1$ and $\alpha = \frac{8}{15}$ results. For these α and β, $\pi = \frac{4}{15}$, but this is easily seen *not* to be a minimal point. In fact, $\pi = \frac{4}{15}$ for all $\beta = -1$ and any α.

The indefinite variational formulation through $\pi(y, M)$ in Eq. (3.58) seems attractive in view of the relaxed continuity requirements on y and M, but it suffers from the decisive deficiency in *convexity*. Because $\pi(y, M)$ does not possess a minimum, $\tilde{M}$ and $\tilde{y}$ do not minimize an error norm and the quality of these approximations is in doubt.

8. A Bound Theorem

An interesting non computational application of the principle of minimum potential energy consists of using it to prove that the solution y to the two point boundary value problem

$$-(p(x)y')' + q(x)y = 0, \qquad 0 < x < 1$$
$$y(0) = y_0, \qquad y(1) = y_1 \tag{3.63}$$

is such that $y_0 \leqslant y \leqslant y_1$. Indeed suppose that $y > y_1$ at some interval $x_0 < x < x_1$ as in Fig. 3.9. The solution y to Eq. (3.63) minimizes

$$\pi(y) = \tfrac{1}{2} \int_0^1 [p(x)y'^2 + q(x)y^2] \, dx \tag{3.64}$$

but the curve y between $x = x_0$ and $x = x_1$ could have been replaced by the horizontal segment, obviously reducing $\pi(y)$, and thereby contradicting the assumption $y > y_1$.

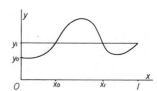

Fig. 3.9 The function y of x rising above y_1 in the interval $x_0 \leqslant x \leqslant x_1$.

EXERCISES

1. Associate with the differential equation for the annular membrane

$$-\frac{1}{r}\frac{d}{dr}\left(r\frac{dy}{dr}\right) + y = f(r), \qquad r_0 < r < r_1$$

the energy norm

$$\|y\|_1{}^2 = \int_{r_0}^{r_1}(y'^2 + y^2)r\,dr$$

and derive the corresponding total potential energy for the boundary conditions $y' = 0$ at $r = r_0$, and $y = 0$ at $r = r_1$.

2.   Modify the energy norm in Exercise 1 to account for the boundary conditions $y = 0$ at $r = r_0$ and $y' + y = 1$ at $r = r_1$. Obtain the total potential energy.

3.   Associate with the pressurized elastic sphere of Exercise 2, Chapter 2, the elastic energy norm

$$\|y\|_1^2 = \int_{r_0}^{r_1} \left(y'^2 + 2\frac{y^2}{r^2} \right) r^2 \, dr$$

and derive the expression for total potential energy for the boundary conditions $y' = -p$ at $r = r_0$ and $y' = 0$ at $r = r_1$.

4.   Associate with the unit circular plate the energy norm

$$\|y\|_2^2 = \int_0^1 \left(y''^2 + \frac{1}{r^2} y'^2 \right) r \, dr$$

and derive the total potential energy

$$\pi(y) = \frac{1}{2} \int_0^1 \left[y''^2 + \frac{1}{r^2} y'^2 - 2y f(r) \right] r \, dr$$

5.   Consider a clamped unit circular plate point loaded at the center with a unit force. Approximate the displacement by $\tilde{y} = \alpha(1 - 3r^2 + 2r^3)$ and minimize $\pi(\tilde{y})$ to obtain the best $\alpha = 1/9$.

6.   For the boundary value problem $-y'' + y = 1$, $y(0) = y(1) = 0$ construct a trial function y in the form of Fig. 3.10: $\tilde{y} = 2\alpha x \; 0 \leqslant x \leqslant \frac{1}{2}$, $\tilde{y} = 2\alpha(1 - x)$, $\frac{1}{2} \leqslant x \leqslant 1$, and obtain the best α by minimizing the relevant $\pi(\tilde{y})$. Compare $y(\frac{1}{2})$ and $\tilde{y}(\frac{1}{2})$.

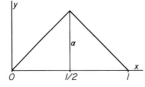

Fig. 3.10   Tent-shaped trial function.

7.   Show that $y = r^{1/2}, 0 \leqslant r \leqslant 1$, is of finite energy for $\|y\|_1$ in Exercise 1.

8.   Show that $y = \log(1/r), 0 \leqslant r \leqslant 1$, is of finite energy for $\|y\|_1$ in Exercise 3.

9.   Show that $y = r^2 \log(1/r), \; 0 \leqslant r \leqslant 1$, is of finite energy for $\|y\|_2$ in Exercise 4.

10. The complementary variational principle of Section 4 can be derived under more relaxed conditions. For this write Eq. (3.12) as $-\sigma' + qy = f$, $y(0) = \sigma(1) = 0$, $\sigma = py'$, and choose the *pair* of trial functions $\bar{y}$ and $\bar{\sigma}$ related by the equilibrium equation $-\bar{\sigma}' + q\bar{y} = f$ and with $\bar{\sigma}(1) = 0$. Start from

$$\frac{1}{2} \int_0^1 \left[p\left(y' - \frac{1}{p}\bar{\sigma} \right)^2 + q(y - \bar{y})^2 \right] dx \geq 0$$

to obtain Eq. (3.43) in the form

$$\psi(\bar{y}, \bar{\sigma}) \leq \pi(y)$$

where

$$\psi(\bar{\sigma}, \bar{y}) = -\frac{1}{2} \int_0^1 \left(\frac{1}{p}\bar{\sigma}^2 + q\bar{y}^2 \right) dx.$$

Write the equation of equilibrium of a cantilever beam as $-M'' + qu = f$, $y(0) = y'(0) = M(1) = M'(1) = \quad$, $M = -py''$, and choose the trial pair $\bar{y}$ and $\bar{M}$ satisfying $-\bar{M}'' + q\bar{y} = f$, $\bar{M}(1) = \bar{M}'(1) = 0$. Starting from

$$\frac{1}{2} \int_0^1 \left[p\left(y'' + \frac{1}{p}\bar{M} \right)^2 + q(y - \bar{y})^2 \right] dx \geq 0$$

show that here

$$\psi(\bar{y}, \bar{M}) = -\frac{1}{2} \int_0^1 \left(\frac{1}{p}\bar{M}^2 + q\bar{y}^2 \right) dx.$$

SUGGESTED FURTHER READING

Sagan, H., *Boundary and Eigenvalue Problems in Mathematical Physics*. Wiley, New York, 1961.
Laughaar, H. L., *Energy Methods in Applied Mechanics*. Wiley, New York, 1962.
Mikhlin, S. G., *Variational Methods in Mathematical Physics*. Pergamon, Oxford, 1964.
Washizu, K., *Variational Methods in Elasticity and Plasticity*. 2nd Ed. Pergamon, Oxford, 1975.

4 *Finite Elements*

1. The Idea of Ritz

In 1908 Walter Ritz had a wonderously simple, but at the same time highly fruitful, idea based on earlier hints by Rayleigh for the numerical solution of boundary value problems that were variationally formulated. It consists of writing the trial function $\tilde{y}$, to be energetically compared with y, as the linear combination

$$\tilde{y} = a_1\phi_1(x) + a_2\phi_2(x) + \cdots + a_N\phi_N(x) \tag{4.1}$$

in which $a_1, a_2, \ldots, a_N$ are weights whose variation controls $\tilde{y}$, and where $\phi_1(x), \phi_2(x), \ldots, \phi_N(x)$ are *basis* or *coordinate functions*. Inasmuch as $\tilde{y}$ is to be introduced into the total potential energy $\pi(\tilde{y})$, the basis functions need be of the proper continuity, they must satisfy the essential boundary conditions that y is required to satisfy, and for reasons of numerical stability to become apparent soon, they must also be linearly independent so that

$$a_1\phi_1(x) + a_2\phi_2(x) + \cdots + a_N\phi_N(x) = 0 \tag{4.2}$$

occurs for all x *only* when all $a_1, a_2, \ldots, a_N$ are zero. To select the *optimal* approximation $\hat{y}$, the one that is energetically closest to y, from among all the possible candidates $\tilde{y}$ in Eq. (4.1), $\tilde{y}$ is introduced into $\pi(\tilde{y})$ and the total potential energy is minimized with respect to $a_1, a_2, \ldots, a_N$ by

$$\frac{\partial \pi}{\partial a_i} = 0, \qquad i = 1, 2, \ldots, N \tag{4.3}$$

furnishing N equations for the N as that determine the best $\tilde{y}$, $\hat{y}$.

Formally, the trial function $\tilde{y}$ in Eq. (4.1) and its first derivative $\tilde{y}'$ are written as

$$\tilde{y} = \sum_{i=1}^{N} a_i \phi_i(x), \qquad \tilde{y}' = \sum_{i=1}^{N} a_i \phi_i'(x) \tag{4.4}$$

or more compactly, with the vector notation $a^{\mathrm{T}} = (a_1, a_2, \ldots, a_N)$, $\phi^{\mathrm{T}} = (\phi_1, \phi_2, \ldots, \phi_N)$ and $\phi'^{\mathrm{T}} = (\phi_1', \phi_2', \ldots, \phi_N')$, $\tilde{y}$ and $\tilde{y}'$ can be written as

$$\tilde{y} = a^{\mathrm{T}}\phi = \phi^{\mathrm{T}}a, \qquad \tilde{y}' = a^{\mathrm{T}}\phi' = \phi'^{\mathrm{T}}a \tag{4.5}$$

When these $\tilde{y}$ and $\tilde{y}'$ are introduced into $\pi(\tilde{y})$, in Eq. (3.20) for example, it becomes

$$\pi(\tilde{y}) = \frac{1}{2} \int_0^1 \left[\sum_{i=1}^{N} \sum_{j=1}^{N} p(x)a_i a_j \phi_i' \phi_j' + \sum_{i=1}^{N} \sum_{j=1}^{N} q(x)a_i a_j \phi_i \phi_j \right.$$
$$\left. - 2 \sum_{i=1}^{N} f(x)a_i \phi_i \right] dx \tag{4.6}$$

or in matrix notation

$$\pi(\tilde{y}) = \frac{1}{2} \int_0^1 \left[p(x)a^{\mathrm{T}}\phi'\phi'^{\mathrm{T}}a + q(x)a^{\mathrm{T}}\phi\phi^{\mathrm{T}}a - 2f(x)a^{\mathrm{T}}\phi \right] dx \tag{4.7}$$

Because a is a constant vector, it can be pulled outside the integral sign to leave $\pi(\tilde{y})$ in the form

$$\pi(\tilde{y}) = \frac{1}{2}a^{\mathrm{T}}\left[\int_0^1 p(x)\phi'\phi'^{\mathrm{T}}dx \right]a + \frac{1}{2}a^{\mathrm{T}}\left[\int_0^1 q(x)\phi\phi^{\mathrm{T}}dx \right]a - a^{\mathrm{T}}\left[\int_0^1 f(x)\phi\, dx \right] \tag{4.8}$$

which can be written concisely as

$$\pi(\tilde{y}) = \tfrac{1}{2}a^{\mathrm{T}}Ka + \tfrac{1}{2}a^{\mathrm{T}}Ma - a^{\mathrm{T}}f \tag{4.9}$$

where the *stiffness* matrix K and the *mass* matrix M are given by

$$K_{ij} = \int_0^1 p(x)\phi_i'\phi_j'\, dx$$
$$M_{ij} = \int_0^1 q(x)\phi_i\phi_j\, dx \tag{4.10}$$

and the *load vector* f by

$$f_i = \int_0^1 f(x)\phi_i\, dx \tag{4.11}$$

Both K and M in Eq. (4.10) are readily seen to be *symmetric*; $K_{ij} = K_{ji}$ and $M_{ij} = M_{ji}$. Moreover, when $q(x) > 0$ the matrix M is also *positive definite* since

$$a^{\mathrm{T}} M a = \int_0^1 q(x) \tilde{y}^2 \, dx > 0 \qquad \text{if} \quad a \neq 0 \tag{4.12}$$

by virtue of the linear independence of the basis functions ϕ. Linear independence among the basis functions causes K to become positive definite too under some easily met restriction on $p(x)$, but this fact is somewhat subtler to prove and we relegate it to Chapter 7.

Minimization of $\pi(\tilde{y})$ in Eq. (4.9) with respect to a^{T} (i.e., $a_1, a_2, \ldots, a_N$) produces the algebraic system of linear equations

$$(K + M)a = f \tag{4.13}$$

that are the ultimate goal of the Ritz method.

As an illustration of Ritz's idea, let us apply his technique to the solution of $-y'' + y = 1$, $y(0) = 0$, $y'(1) = 0$, that we already solved in Chapter 3, Section 4, with only one basis function. This time we take

$$\tilde{y} = a_1 x + a_2 x^2 + a_3 x^3 \tag{4.14}$$

in which $\phi^{\mathrm{T}} = (x, x^2, x^3)$ and $\phi'^{\mathrm{T}} = (1, 2x, 3x^2)$. As required, the basis functions ϕ, and also ϕ', are linearly independent and $\phi_1(0) = \phi_2(0) = \phi_3(0) = 0$, making $\tilde{y}(0)$ satisfy the essential boundary condition $y(0) = 0$. Emerging from the general procedure of this section, with $\tilde{y}$ in Eq. (4.14), is the algebraic system

$$\left\{ \frac{1}{30} \begin{bmatrix} 30 & 30 & 30 \\ 30 & 40 & 45 \\ 30 & 45 & 54 \end{bmatrix} + \frac{1}{420} \begin{bmatrix} 140 & 105 & 84 \\ 105 & 84 & 70 \\ 84 & 70 & 60 \end{bmatrix} \right\} \begin{bmatrix} a_1 \\ a_2 \\ a_3 \end{bmatrix} = \frac{1}{12} \begin{bmatrix} 6 \\ 4 \\ 3 \end{bmatrix} \tag{4.15}$$

in which both K and M are symmetric, positive definite, and hence nonsingular, and the system can be numerically solved.

The motive for including more and more basis functions in the expression (4.1) for $\tilde{y}$ is the hope that with more basis functions a better approximation $\tilde{y}$ to y will become possible which will be picked up by the minimization of the total potential energy. As presented up to this point, the method of Ritz does not include a provision for the *systematic* inclusion of *good* basis functions. Physical intuition and previous experience could provide hints to a judicious choice of basis functions, such that only few of them would be sufficient for acceptable accuracy. On the other hand, if the solution y is misfigured, as can easily happen when it is complicated and discontinuous, an enormous number of basis functions might be required to adequately approximate this solution. As the number of basis functions increase, the algebraic system formed by the Ritz method increases in size, becoming more expensive to solve. Not only this, but the matrices associated with this linear system of equations might also decline in condition to the point of, defying the computer accuracy, becoming numerically singular. That this is not a farfetched eventuality is seen from the numerical example that produced Eq. (4.15). Both the stiffness matrix K and the mass (actually *Gram*) matrix M derived from the polynomial basis $\phi^T = (1, x, x^2, \dots)$ are the notoriously ill conditioned *Hilbert* matrices. These innocent looking matrices turn ill conditioned so fast that the spectral condition number of the 3×3 matrix is already 270. For $N = 4$, 5, and 6 it grows exponentially to 6800, 20,000, and 800,000, respectively, and the 6×6 K matrix is practically indefinite on an 8 digit computer.

Finite elements eliminate these two difficulties of the Ritz method. The method provides an automatic technique for generating a converging sequence of trial functions $\tilde{y}$, ever diminishing $\|y - \hat{y}\|_m$ ($m = 1$ for the string problem and $m = 2$ for the beam problem), and at the same time the matrices set up by this method are, for acceptable accuracy, sufficiently well conditioned for a comfortable round-off error margin on existing computers that typically carry from 6 to 32 significant digits.

2. Finite Element Basis Functions

The underlying idea of finite elements is the construction of a trial function $\tilde{y}$ for the Ritz method, consisting of polynomial segments that need meet at the joints only the bare minimal continuity requirements for the use of $\tilde{y}$ in the principle of minimum potential energy. Second-order problems, those with no higher derivatives than y' in $\pi(y)$, require no more than C^0 continuity in $\tilde{y}$ and the piecewise polynomial $\tilde{y}$ can even be piecewise linear. The advantages of this device for constructing $\tilde{y}$, coupled with the variational formulation of the boundary value problem, are many. For one thing, it relieves us of the need to care for the approximation of the higher-order natural boundary conditions. For the other, the matrices produced by finite elements are, like those formed by the Ritz method, symmetric, essential boundary conditions are enforced in an unambiguous manner, variation of the mesh size or change of the polynomial interpolation order is routinely done to better control the accuracy, graded meshes to meet local approximation requirements are as simple to program as a uniform mesh is, discontinuities are readily duplicated, and numerical stability, as we shall later see, is guaranteed with finite elements.

Construction of the piecewise linear finite element trial function $\tilde{y}$ reminds
us of finite differences. As with finite differences here also the x interval is
subdivided, say uniformly, into Ne segments, or *finite elements*, each of size h.
The joints of the finite elements are the *nodal points*, labeled from 1 to N in
the way shown in Fig. 4.1. Each node is assigned a *nodal value*—the value
of y or $\tilde{y}$ at this node, which is subscripted accordingly. A linear interpolation
of the nodal values inside each element creates a continuous, piecewise
linear, trial function $\tilde{y}$ as shown in Fig. 4.1. Changing the nodal values
$y_1, y_2, \ldots, y_5$ in Fig. 4.1 causes the configuration of $\tilde{y}$ to change and in this
way the nodal values play the role of the as in Eq. (4.1). But, while the as
in the Ritz method are not of immediate interest, the finite element weights
or nodal values are of direct interest.

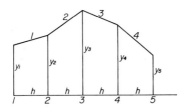

Fig. 4.1 Piecewise linear displacement trial function over four finite elements.

All the $\tilde{y}$ choices formed with different nodal values $y_1, y_2, \ldots, y_N$ of the
same finite element position are said, in the terminology of functional analysis, to lie in a finite dimensional function space $\tilde{y}$. This space is spanned by
some N basis functions $\phi_1, \phi_2, \ldots, \phi_N$. A basis for a certain space is not
unique and various linear combinations of the old basis result in other bases;
all in analogy with vector spaces.

To discover what the basis or coordinate functions are for $\tilde{y}$ in Fig. 4.1,
we write $\tilde{y} = y_1\phi_1 + y_2\phi_2 + \cdots + y_N\phi_N$ and observe that when $y_j = 1$, and
all other ys are zero, $\tilde{y} = \phi_j$. Five tent-shaped basis functions are obtained
from this and are displayed in Fig. 4.2. Because ϕ_j is nonzero only over a small
interval $2h$, it is said to have a *compact support*. The compact support causes
the mass and stiffness matrices formed with these basis functions to become
sparse. In fact, K and M, formed with $\phi_1, \phi_2, \ldots, \phi_N$ of Fig. 4.2, are *tridiagonal*
since

$$K_{ij} = \int_0^1 p(x)\phi_i'\phi_j' = 0, \qquad M_{ij} = \int_0^1 q(x)\phi_i\phi_j \, dx = 0, \qquad |i - j| > 1$$
$$(4.16)$$

Two basis functions whose support does not overlap are *orthogonal* with
respect to any weight function.

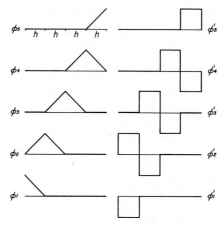

Fig. 4.2 Five tentlike basis function $\phi_1, \phi_2, \ldots, \phi_5$ and their derivatives for the piecewise linear trial function of Fig. 4.1 when $y_1, y_2, \ldots, y_5$ are the Ritz weights.

Once the finite element basis functions are determined, we may proceed to form with them the Ritz linear system. Finite elements do, however, more than merely provide basis functions for the Ritz technique. They embody also a shrewd, repetitive, and remarkably easy to program technique for the assembly of the stiffness and mass matrices that bypass the basis functions and operate instead on a succession of, actually identical, finite elements.

3. Finite Element Matrices

Computation of the total potential energy is carried out in the finite element method through the summation of $\pi(\bar{y})$, actually computed *only once over a typical element*. To systematize the computation of $\pi(\bar{y})$ over a single element, the nodes of the typical element shown in Fig. 4.3, are labeled 1, 2, which constitutes the *element nodal numbering system*, as opposed to the *global* system of nodal numbering in Fig. 4.1. Each element is thus associated

Fig. 4.3 A typical linear finite element.

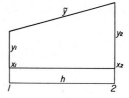

with a y_1 and y_2, referring to the left- and right-hand points 1 and 2. Simplicity in the computation of $\pi(\tilde{y})$ is gained when the typical finite element is mapped from $x_1 \leqslant x \leqslant x_1 + h = x_2$ to the interval $0 \leqslant \xi \leqslant 1$ by

$$x = x_1 + h\xi \tag{4.17}$$

In terms of ξ, the linear interpolation of $\tilde{y}$ inside the element is made by

$$\tilde{y} = y_1\phi_1(\xi) + y_2\phi_2(\xi)$$
$$\phi_1(\xi) = 1 - \xi, \phi_2(\xi) = \xi \tag{4.18}$$

where $\phi_1(\xi)$ and $\phi_2(\xi)$ are the element *shape functions*.

Let $\pi_e(\tilde{y})$ denote the total potential energy for one element. Then in terms of ξ, the total potential energy of Eq. (3.20) may be written as

$$\pi_e(\tilde{y}) = \frac{1}{2} \int_0^1 \left[\frac{1}{h^2} p(\xi) \left(\frac{dy}{d\xi} \right)^2 + q(\xi)y^2 - 2f(\xi)y \right] h \, d\xi \tag{4.19}$$

Substitution of $\tilde{y}$ into this $\pi_e(\tilde{y})$ is entirely analogous to the procedure of Section 1 for the Ritz method, except that here only one typical element is considered. Corresponding to Eq. (4.5), we write

$$\tilde{y} = y_e^T\phi = \phi^Ty_e, \qquad \tilde{y}' = y_e^T\phi' = \phi'^Ty_e \tag{4.20}$$

in which ϕ is the shape functions vector, and y_e the vector of the nodal values of the element. Upon the introduction of $\tilde{y}$ in Eq. (4.20) into $\pi_e(\tilde{y})$ in Eq. (4.19), the latter becomes

$$\pi_e(\tilde{y}) = \frac{1}{2} \int_0^1 \left[\frac{1}{h} p(\xi)y_e^T\phi'\phi'^Ty_e + hq(\xi)y_e^T\phi\phi^Ty_e - 2hf(\xi)y_e^T\phi \right] d\xi \tag{4.21}$$

In short

$$\pi_e(\tilde{y}) = \frac{1}{2}y_e^T(k_e + m_e)y_e - y_e^Tf_e \tag{4.22}$$

where k_e is the *element stiffness matrix*, m_e the *element mass matrix*, and f_e the *element load vector*; a terminology borrowed from structural mechanics where finite elements originated, and where an element may be a real structural member, or element, of some stiffness and mass. The physically inclined may consider the piecewise linear approximation of the string as the real replacement of the string by pin-joined rods.

Equations (4.21) and (4.22) are good for any shape functions vector, and in general

$$k_e = \frac{1}{h} \int_0^1 p(\xi)\phi'\phi'^T \, d\xi, \qquad m_e = h \int_0^1 q(\xi)\phi\phi^T \, d\xi,$$
$$f_e = h \int_0^1 f(\xi)\phi \, d\xi \tag{4.23}$$

In the particular case of a linear piecewise $\tilde{y}$, where $\phi^T = (1 - \xi, \xi)$ and $\phi'^T = (-1, 1)$, and in the special case of $p(x) = q(x) = f(x) = 1$, the element matrices become

$$k_e = \frac{1}{h}\begin{bmatrix} 1 & -1 \\ -1 & 1 \end{bmatrix}, \quad m_e = \frac{h}{6}\begin{bmatrix} 2 & 1 \\ 1 & 2 \end{bmatrix}, \quad f_e = \frac{h}{2}\begin{bmatrix} 1 \\ 1 \end{bmatrix} \tag{4.24}$$

Since the derivation of the element matrices is restricted to one element at a time, it is immaterial whether the mesh size h is constant or not. If the mesh is graded, h in Eq. (2.24) need merely change from element to element. When $p(x)$, $q(x)$, and $f(x)$ depend on x, the integration of $\pi_e(\tilde{y})$ for the formation of k_e, m_e, and f_e becomes more involved and numerical integration is in order. More on this later, in Chapter 5, Section 7.

4. Assembly of Global Matrices

For the systematic description of the summation of $\pi(\tilde{y})$ from $\pi_e(\tilde{y})$, we introduce the *global* vector $y^T = (y_1, y_2, \ldots, y_N)$ of nodal values, arranged relative to the global numbering system. This global vector is related to the element nodal values vector y_e by

$$y_e = A_e y \tag{4.25}$$

where A_e are *Boolean* mapping matrices containing only 1s and 0s, with the letter A chosen for Argyris who first introduced this mapping. In the particular case of Fig. 4.1, the four A_e matrices for the four linear finite elements are

global numbering
system

1 2 3 4 5

$$A_1 = \begin{bmatrix} 1 & & & & \\ & 1 & & & \end{bmatrix} \begin{matrix} 1 \\ 2 \end{matrix} \begin{matrix} \text{local numbering} \\ \text{system} \end{matrix}, \quad A_2 = \begin{bmatrix} & 1 & & & \\ & & 1 & & \end{bmatrix}$$

$$A_3 = \begin{bmatrix} & & 1 & & \\ & & & 1 & \end{bmatrix}, \quad A_4 = \begin{bmatrix} & & & 1 & \\ & & & & 1 \end{bmatrix} \tag{4.26}$$

Of course $(A_e)_{ij} = 1$ when the local number i corresponds to the global j; otherwise $(A_e)_{ij} = 0$. These sparse matrices are introduced only for notational convenience and are excluded from the actual computational procedure of setting up K, M, and f.

With equation (4.25), $\pi_e(\tilde{y})$ is written as

$$\pi_e(\tilde{y}) = \tfrac{1}{2}y^{\mathrm{T}}(A_e^{\mathrm{T}}k_eA_e + A_e^{\mathrm{T}}m_eA_e)y - y^{\mathrm{T}}A_e^{\mathrm{T}}f_e \qquad (4.27)$$

from which the sum

$$\pi(\tilde{y}) = \tfrac{1}{2}y^{\mathrm{T}}\left[\sum_{e=1}^{Ne} (A_e^{\mathrm{T}}k_eA_e + A_e^{\mathrm{T}}m_eA_e)\right]y - y^{\mathrm{T}}\sum_{e=1}^{Ne} A_e^{\mathrm{T}}f_e \qquad (4.28)$$

is obtained. Or in short,

$$\pi(\tilde{y}) = \tfrac{1}{2}y^{\mathrm{T}}(K + M)y - y^{\mathrm{T}}f \qquad (4.29)$$

where

$$K = \sum_{e=1}^{Ne} A_e^{\mathrm{T}}k_eA_e, \qquad M = \sum_{e=1}^{Ne} A_e^{\mathrm{T}}m_eA_e, \qquad f = \sum_{e=1}^{Ne} A_e^{\mathrm{T}}f_e \qquad (4.30)$$

are the *global stiffness* and *mass matrices* and the *global load vector*. Minimization of $\pi(\tilde{y})$ in Eq. (4.29) with respect to y^{T} produces the linear algebraic system $(K + M)y = f$ which was also the goal of the Ritz method.

The assembly of K, M, and f from the corresponding element data is utterly simple to execute on the computer. At the heart of this procedure lies the *congruent transformation* $A_e^{\mathrm{T}}k_eA_e$ which means the following: If k_e is $n \times n$ and K is $N \times N$, then k_e is enlarged from $n \times n$ into a matrix $K_e = A_e^{\mathrm{T}}k_eA_e$ which is $N \times N$, with the entry $(k_e)_{ij}$ going into the entry $(K_e)_{mn}$, the local pair of nodal numbers i, j corresponding to the global pair m, n. As we said before, the Argyris matrices A_e are introduced only to facilitate the symbolic description of the summation of $\pi_e(\tilde{y})$ into $\pi(\tilde{y})$; they do not enter as such into the computer algorithm. All that is needed for carrying out the summation of k_e, m_e, and f_e into K, M, and f is the correspondence, for each element, between its local and global numbering systems that may be collected into a single table like

1	2	local numbering system

element 1	1	2	
element 2	2	3	
element 3	3	4	global numbering system
element 4	4	5	

for the mesh in Fig. 4.1. According to this table,

$$A_1{}^T k_1 A_1 = K_1 = \frac{1}{h} \begin{bmatrix} 1 & -1 & & \\ -1 & 1 & & \\ & & & \\ & & & \end{bmatrix}, \qquad A_2{}^T k_2 A_2 = K_2 = \frac{1}{h} \begin{bmatrix} & & & \\ & 1 & -1 & \\ & -1 & 1 & \\ & & & \end{bmatrix}$$

$$A_3{}^T k_3 A_3 = K_3 = \frac{1}{h} \begin{bmatrix} & & & \\ & & 1 & -1 \\ & & -1 & 1 \\ & & & \end{bmatrix}, \qquad A_4{}^T k_4 A_4 = K_4 = \frac{1}{h} \begin{bmatrix} & & & \\ & & & \\ & & 1 & -1 \\ & & -1 & 1 \end{bmatrix}$$

$$A_1{}^T m_1 A_1 = M_1 = \frac{h}{6} \begin{bmatrix} 2 & 1 & & \\ 1 & 2 & & \\ & & & \\ & & & \end{bmatrix}, \qquad A_2{}^T m_2 A_2 = M_2 = \frac{h}{6} \begin{bmatrix} & & & \\ & 2 & 1 & \\ & 1 & 2 & \\ & & & \end{bmatrix}$$

$$A_3{}^T m_3 A_3 = M_3 = \frac{h}{6} \begin{bmatrix} & & & \\ & & 2 & 1 \\ & & 1 & 2 \\ & & & \end{bmatrix}, \qquad A_4{}^T m_4 A_4 = M_4 = \frac{h}{6} \begin{bmatrix} & & & \\ & & & \\ & & 2 & 1 \\ & & 1 & 2 \end{bmatrix}$$

$$A_1{}^T f_1 = \frac{h}{2} \begin{bmatrix} 1 \\ 1 \\ \\ \end{bmatrix}, \qquad A_2{}^T f_2 = \frac{h}{2} \begin{bmatrix} \\ 1 \\ 1 \\ \end{bmatrix},$$

$$A_3{}^T f_3 = \frac{h}{2} \begin{bmatrix} \\ 1 \\ 1 \\ \end{bmatrix}, \qquad A_4{}^T f_4 = \frac{h}{2} \begin{bmatrix} \\ \\ 1 \\ 1 \end{bmatrix}$$

in which a typical $A_e^T k_e A_e$ and $A_e^T f_e$ is constructed by

$$A^T k A = \begin{bmatrix} & \overset{m}{k_{11}} & \overset{n}{k_{12}} \\ & k_{21} & k_{22} \end{bmatrix} \begin{matrix} m, \\ n \end{matrix} \qquad A^T f = \begin{bmatrix} f_1 \\ f_2 \end{bmatrix} \begin{matrix} m \\ n \end{matrix} \qquad (4.31)$$

where the global pair m, n corresponds to the local 1, 2. Formation of all $A_e^T f_e$ and their summation into the global f should be clear from Eq. (4.31). After all—here four, $A_e^T k_e A_e$, $A_e^T m_e A_e$, and $A_e^T f_e$ have been formed and summed over the four Ne finite elements the matrices

$$K = \frac{1}{h} \begin{bmatrix} 1 & -1 & & & \\ -1 & 2 & -1 & & \\ & -1 & 2 & -1 & \\ & & -1 & 2 & -1 \\ & & & -1 & 1 \end{bmatrix}, \quad M = \frac{h}{6} \begin{bmatrix} 2 & 1 & & & \\ 1 & 4 & 1 & & \\ & 1 & 4 & 1 & \\ & & 1 & 4 & 1 \\ & & & 1 & 2 \end{bmatrix}, \quad f = \frac{h}{2} \begin{bmatrix} 1 \\ 2 \\ 2 \\ 2 \\ 1 \end{bmatrix}$$

(4.32)

result. Notice that K in Eq. (4.32) contains the coefficients of the central difference scheme for y''. Starting from a variational formulation finite elements generates finite difference equations, albeit in a remarkably systematic fashion.

A FORTRAN program would carry out the assembly of K, M, and f in the following way. Let $sk(n, n)$, $sm(n, n)$, and $sf(n)$ be the small element matrices. Let $gk(N, N)$, $gm(N, N)$, and $gf(N)$ be their global counterpart. Also, let $ng(Ne, n)$ be the list of global numbers for the Ne finite elements. Then

$$\begin{aligned}
&\text{do 1 } ii = 1, ne \\
&\text{do 1 } i = 1, n \\
&k = ng(ii, i) \\
&gf(k) = gf(k) + sf(i) \\
&\text{do 1 } j = 1, n \\
&l = ng(ii, j) \\
&gk(k, l) = gk(k, l) + sk(i, j) \\
&\text{1 } gm(k, l) = gm(k, l) + sm(i, j)
\end{aligned}$$

where ii ranges over all $Ne \equiv ne$ finite elements, and where i and j are the local or element members to which the global k and l correspond.

5. Essential and Natural Boundary Conditions

The assembly procedure for K, M, and f described in the previous section does not include the enforcement of the essential boundary conditions $\tilde{y}$ need satisfy. Because of their variational origin, finite elements do not require the *a priori* fulfillment of the variationally natural boundary conditions. To describe the enforcement of the essential boundary conditions, assume first that those are only $\tilde{y}(0) = y_1 = 0$. Variation of $\pi(\tilde{y})$ with respect to y_1 is, in this case, identically zero, meaning that the first equation in $(K + M)y = f$ should be deleted. Moreover, since $y_1 = 0$, all terms in $\pi(\tilde{y})$ that include y_1 vanish, which is equivalent to the vanishing of the first *column* in K and M. Thus, the introduction of the essential boundary condition $y_1 = 0$ is effected by the deletion of both the first row and columns in $K + M$ and the first entry in f, reducing thereby the size of the linear algebraic system from $N \times N$ to $(N - 1) \times (N - 1)$. Similarly if $y_N = 0$, the Nth row and column are deleted. Removal of rows and columns of two-dimensional arrays and their compression is, however, inconvenient with the present-day computer programming languages and it would be better to leave the matrices at their original dimension. To avoid singularity the equation $y_1 = 0$ is added to the algebraic system by putting a one for $(K + M)_{11}$ and a zero at f_1. If $y_1 = 0$ is the only essential boundary condition, the algebraic system with the matrices of Eq. (4.32) becomes

$$\left\{ \frac{1}{h}\begin{bmatrix} 1 & & & \\ -1 & 2 & -1 & \\ & -1 & 2 & -1 \\ & & -1 & 1 \end{bmatrix} + \frac{h}{6}\begin{bmatrix} 4 & 1 & & \\ 1 & 4 & 1 & \\ & 1 & 4 & 1 \\ & & 1 & 2 \end{bmatrix} \right\} y = \frac{h}{2}\begin{bmatrix} 2 \\ 2 \\ 2 \\ 1 \end{bmatrix} \qquad (4.33)$$

Nonhomogeneous essential boundary conditions are introduced by partitioning the global matrices and vectors into a portion corresponding to the prescribed values and a portion corresponding to the free variables. For $y_1 = y_a \neq 0$ this partitioning is of the form

$$K = \begin{bmatrix} K_{11} & | & K_{12} \\ \hline K_{12}^T & | & K_{22} \end{bmatrix}, \quad M = \begin{bmatrix} M_{11} & | & M_{12} \\ \hline M_{12}^T & | & M_{22} \end{bmatrix}, \quad f = \begin{bmatrix} f_1 \\ \hline f_2 \end{bmatrix}, \quad y = \begin{bmatrix} y_1 \\ \hline y_2 \end{bmatrix}$$

$$(4.34)$$

where y_1 is the prescribed value, or even group of values, and y_2 the rest. In its partitioned form $\pi(\tilde{y})$ is

$$\pi(\tilde{y}) = \tfrac{1}{2}y_1{}^T(K_{11} + M_{11})y_1 + \tfrac{1}{2}y_2{}^T(K_{22} + M_{22})y_2$$
$$+ y_2{}^T(K_{12}^T + M_{12}^T)y_1 - y_1{}^Tf_1 - y_2{}^Tf_2 \qquad (4.35)$$

and since y_1 is kept fixed and $\pi(\tilde{y})$ varied only with respect to y_2, $\partial\pi(\tilde{y})/\partial y_2 = 0$ yields

$$(K_{22} + M_{22})y_2 = f_2 - (K_{12} + M_{12})^T y_1 \tag{4.36}$$

Again, the essential boundary conditions are introduced by deleting the corresponding rows and columns of K, M, and f, except that now $-(K_{12} + M_{12})^T y_1$ is added to the right-hand side of Eq. (4.36). Reduction of size in the algebraic system is avoided by replacing the deleted equations with the boundary conditions. For the case $y(0) = y_1 = y_a$ in our example, the linear system becomes

$$\left\{\frac{1}{h}\begin{bmatrix} h^2/2 & & & & \\ & 2 & -1 & & \\ & -1 & 2 & -1 & \\ & & -1 & 2 & -1 \\ & & & -1 & 2 \end{bmatrix} + \frac{h}{6}\begin{bmatrix} & & & & \\ & 4 & 1 & & \\ & 1 & 4 & 1 & \\ & & 1 & 4 & 1 \\ & & & 1 & 4 \end{bmatrix}\right\}\begin{bmatrix} y_1 \\ y_2 \\ y_3 \\ y_4 \\ y_5 \end{bmatrix}$$

$$= \frac{h}{2}\begin{bmatrix} y_a \\ 2 \\ 2 \\ 2 \\ 1 \end{bmatrix} - y_a\left\{\frac{1}{h}\begin{bmatrix} -1 \\ \\ \\ \\ \end{bmatrix} + \frac{h}{6}\begin{bmatrix} 1 \\ \\ \\ \\ \end{bmatrix}\right\} \tag{4.37}$$

Natural boundary conditions that change $\pi(y)$, as in Chapter 3, Section 3, where boundary terms are added to the total potential energy, are simply incorporated by adding h_0 to $(K + M)_{11}$, h_1 to $(K + M)_{55}$, and c_0 and c_1 to f_1 and f_5, respectively. The linear system resulting is now

$$\left\{\frac{1}{h}\begin{bmatrix} 1 + hh_0 & -1 & & & \\ -1 & 2 & -1 & & \\ & -1 & 2 & -1 & \\ & & -1 & 2 & -1 \\ & & & -1 & 1 + hh_1 \end{bmatrix} + \frac{h}{6}\begin{bmatrix} 2 & 1 & & & \\ 1 & 4 & 1 & & \\ & 1 & 4 & 1 & \\ & & 1 & 4 & 1 \\ & & & 1 & 2 \end{bmatrix}\right\}y$$

$$= \frac{h}{2}\begin{bmatrix} 1 + 2c_0/h \\ 2 \\ 2 \\ 2 \\ 1 + 2c_1/h \end{bmatrix} \tag{4.38}$$

Unlike finite differences, the introduction of the natural, as well as the essential, boundary conditions in the finite element method is systematic and unequivocal.

6. Higher-Order Finite Elements

One of the advantages of the finite element trial function, over that of Ritz, is that the former can easily be visualized in its dependence upon the nodal values, unlike the latter where the variation of the weights cause $\tilde{y}$ to change in a manner hard to imagine. Visualization is probably best with linear elements but a better approximation, with the same element size, seems to be attainable with a higher-order polynomial interpolation of $\tilde{y}$ inside the element. A piecewise *quadratic* $\tilde{y}$ over three elements, each of size $2h$, is shown in Fig. 4.4. Now each element must have three nodal points for the parabolic $\tilde{y}$ inside it: one at each end to ensure C^0 continuity and one we put in the center. A typical quadratic element is shown in Fig. 4.5. Simplicity in the analysis is gained by mapping the typical element to the interval $-1 \leqslant \xi \leqslant 1$ by the transformation

$$x = x_2 + h\xi, \qquad -1 \leqslant \xi \leqslant 1 \tag{4.39}$$

and $\tilde{y}$ inside the element is written in terms of the *three* shape functions $\phi_1(\xi), \phi_2(\xi), \phi_3(\xi)$ as

$$\tilde{y} = y_1\phi_1(\xi) + y_2\phi_2(\xi) + y_3\phi_3(\xi) \tag{4.40}$$

Fig 4.4 Piecewise quadratic trial function over three finite elements.

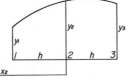

Fig. 4.5 A typical quadratic finite element.

From the conditions that $\phi_j(\xi)$, $j = 1, 2, 3$, is parabolic, $\phi_j(\xi) = 1$ at the jth node, and $\phi_j(\xi) = 0$ at all other nodes we get that

$$\phi_1(\xi) = \tfrac{1}{2}\xi(\xi - 1), \qquad \phi_2(\xi) = 1 - \xi^2, \qquad \phi_3(\xi) = \tfrac{1}{2}\xi(\xi + 1) \tag{4.41}$$

Indeed, all three are quadratic: $\phi_1 = 1$ at nodal point 1, $\phi_1 = 0$ at nodes 2 and 3, and so on, as graphically shown in Fig. 4.6. Introduction of $\phi^{\mathrm{T}} = (\phi_1, \phi_2, \phi_3)$ into Eq. (4.23) results, with $p(x) = q(x) = f(x) = 1$, in

$$k_{\mathrm{e}} = \frac{1}{6h}\begin{bmatrix} 7 & -8 & 1 \\ -8 & 16 & -8 \\ 1 & -8 & 7 \end{bmatrix}, \qquad m_{\mathrm{e}} = \frac{h}{15}\begin{bmatrix} 4 & 2 & -1 \\ 2 & 16 & 2 \\ -1 & 2 & 4 \end{bmatrix}, \qquad f_{\mathrm{e}} = \frac{h}{3}\begin{bmatrix} 1 \\ 4 \\ 1 \end{bmatrix} \tag{4.42}$$

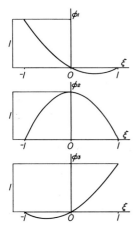

Fig. 4.6   Shape functions of the quadratic element in Fig. 4.5.

To the mesh in Fig. 4.4, corresponds the global numbering list

local

		1	2	3
	1	1	2	3
element	2	3	4	5
	3	5	6	7

from which the global matrices

$$K = \frac{1}{6h} \begin{bmatrix} 7 & -8 & 1 & & & & \\ -8 & 16 & -8 & & & & \\ 1 & -8 & 14 & -8 & 1 & & \\ & & -8 & 16 & -8 & & \\ & & 1 & -8 & 14 & -8 & 1 \\ & & & & -8 & 16 & -8 \\ & & & & 1 & -8 & 7 \end{bmatrix},$$

$$M = \frac{h}{15} \begin{bmatrix} 4 & 2 & -1 & & & & \\ 2 & 16 & 2 & & & & \\ -1 & 2 & 8 & 2 & -1 & & \\ & & 2 & 16 & 2 & & \\ & & -1 & 2 & 8 & 2 & -1 \\ & & & & 2 & 16 & 2 \\ & & & & -1 & 2 & 4 \end{bmatrix}, \quad f = \frac{h}{3} \begin{bmatrix} 1 \\ 4 \\ 2 \\ 4 \\ 2 \\ 4 \\ 1 \end{bmatrix} \qquad (4.43)$$

are constructed, and later modified to accomodate the prevailing boundary conditions.

Often, the derivative y' is of physical significance and we are interested in including it among the nodal values to form a trial function $\tilde{y}$ that is not only C^0 but also C^1, as shown in Fig. 4.7. A typical element in the mesh shown in Fig. 4.7 is separately shown in Fig. 4.8. Each of these elements is associated with only two nodal points but with the four nodal values y_1, y_1', y_2, y_2' that determine a cubic $\tilde{y}$ inside the element. To find out the shape functions, we write $\tilde{y}$ inside the element in the form

$$\tilde{y}(\xi) = a_0 + a_1\xi + a_2\xi^2 + a_3\xi^3, \qquad 0 \leqslant \xi \leqslant 1 \qquad (4.44)$$

and express a_0, a_1, a_2, and a_3 in terms of y_1, y_1', y_2, and y_2' from the conditions $\tilde{y}(0) = y_1$, $\tilde{y}(1) = y_2$, $\tilde{y}'(0) = y_1'$, and $\tilde{y}'(1) = y_2'$, where y' means dy/dx. Then with the arrangement

$$\tilde{y} = y_1\phi_1 + y_1'\phi_2 + y_2\phi_3 + y_2'\phi_4 \qquad (4.45)$$

we have that

$$\phi_1 = 1 - 3\xi^2 + 2\xi^3, \qquad \phi_2 = h(\xi - 2\xi^2 + \xi^3),$$
$$\phi_3 = 3\xi^2 - 2\xi^3, \qquad \phi_4 = h(-\xi^2 + \xi^3) \qquad (4.46)$$

that are traced in Fig. 4.9.

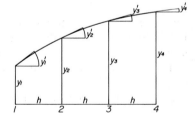

Fig. 4.7 A C^1 trial function over three elements. Both y and y' are continuous at the nodes.

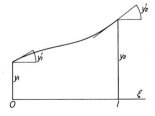

Fig. 4.8 A cubic element with displacements and slopes as nodal values.

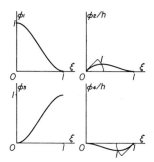

Fig. 4.9   Four shape functions of the element in Fig. 4.8.

Once the shape functions are known for the element, the formation of k_e, m_e, and f_e follows from Eq. (4.23). Here, with $p(x) = q(x) = f(x) = 1$,

$$k_e = \frac{1}{30h} \begin{bmatrix} 36 & 3h & -36 & 3h \\ 3h & 4h^2 & -3h & -h^2 \\ -36 & -3h & 36 & -3h \\ 3h & -h^2 & -3h & 4h^2 \end{bmatrix},$$

$$m_e = \frac{h}{420} \begin{bmatrix} 156 & 22h & 54 & -13h \\ 22h & 4h^2 & 13h & -3h^2 \\ 54 & 13h & 156 & -22h \\ -13h & -3h^2 & -22h & 4h^2 \end{bmatrix}, \qquad f_e = \frac{h}{12} \begin{bmatrix} 6 \\ h \\ 6 \\ -h \end{bmatrix} \qquad (4.47)$$

Taking $\theta = hy'$ as a nodal value instead of y' removes h from the shape functions in Eq. (4.46) and consequently also from the interior of k_e, m_e, and f_e.

The complete list of nodal values for the three element mesh of Fig. 4.8 is $(y_1, \theta_1, y_2, \theta_2, y_3, \theta_3, y_4, \theta_4)$, and we number them globally from 1 to 8. Relative to this nodal values listing is the local–global numbering table

		local			
		1	2	3	4
	1	1	2	3	4
element	2	3	4	5	6
	3	5	6	7	8

from which the following K, M, and f are assembled

$$K = \frac{1}{30h} \begin{bmatrix} 36 & 3 & -36 & 3 & & & & \\ 3 & 4 & -3 & -1 & & & & \\ -36 & -3 & 72 & & -36 & 3 & & \\ 3 & -1 & & 8 & -3 & -1 & & \\ & & -36 & -3 & 72 & & -36 & 3. \\ & & 3 & -1 & & 8 & -3 & -1 \\ & & & & -36 & -3 & 36 & -3 \\ & & & & 3 & -1 & -3 & 4 \end{bmatrix},$$

$$M = \frac{h}{420} \begin{bmatrix} 156 & 22 & 54 & -13 & & & & \\ 22 & 4 & 13 & -3 & & & & \\ 54 & 13 & 312 & & 54 & -13 & & \\ -13 & -3 & & 8 & 13 & -3 & & \\ & & 54 & 13 & 312 & & 54 & -13 \\ & & -13 & -3 & & 8 & 13 & -3 \\ & & & & 54 & 13 & 156 & -22 \\ & & & & -13 & -3 & -22 & 4 \end{bmatrix},$$

$$f = \frac{h}{12} \begin{bmatrix} 6 \\ 1 \\ 12 \\ 12 \\ 12 \\ 6 \\ -1 \end{bmatrix} \qquad (4.48)$$

In case the mesh of finite elements is not uniform and includes elements of different size, the transformation $\theta = hy'$ has to be made with some average h, since otherwise θ from the left element might be different, on account of the different element size, from that to the right of the nodal point.

A C^1 continuity in $\tilde{y}$ is not required in second-order problems and the last element may be assembled without matching θ at the nodal points. This leaves each interior nodal point with three nodal values θ^-, y, θ^+ as shown in Fig. 4.10, and the complete list of nodal values for the three element mesh

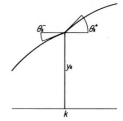

Fig. 4.10 Trial function with the slope θ remaining unconnected at the node.

is $(y_1, \theta_1, \theta_2^-, y_2, \theta_2^+, \theta_3^-, y_3, \theta_3^+, y_4, \theta_4)$, numbered globally from 1 to 10 and with the local–global relationship

local

		1	2	3	4
	1	1	2	4	3
element	2	4	5	7	6
	3	7	8	9	10

from which

$$K = \frac{1}{30h}\begin{bmatrix} 36 & 3 & 3 & -36 & & & & & & \\ 3 & 4 & -1 & -3 & & & & & & \\ 3 & -1 & 4 & -3 & & & & & & \\ -36 & -3 & -3 & 72 & 3 & 3 & -36 & & & \\ & & & 3 & 4 & -1 & -3 & & & \\ & & & 3 & -1 & 4 & -3 & & & \\ & & & -36 & -3 & -3 & 72 & 3 & -36 & 3 \\ & & & & & & 3 & 4 & -3 & -1 \\ & & & & & & -36 & -3 & 36 & -3 \\ & & & & & & 3 & -1 & -3 & 4 \end{bmatrix} \quad (4.49)$$

is assembled.

If we have in mind to assemble the previous element without connecting y' at the nodes, we had better change the element nodal values sequence to (y_1, y_1', y_2', y_2) with the corresponding k_e:

$$k_e = \frac{1}{30h}\begin{bmatrix} 36 & 3h & 3h & -36 \\ 3h & 4h^2 & -h^2 & -3h \\ 3h & -h^2 & 4h^2 & -3h \\ -36 & -3h & -3h & 36 \end{bmatrix} \quad (4.50)$$

Related to this element nodal value arrangement is the simpler local–global correspondance

local

		1	2	3	4
	1	1	2	3	4
element	2	4	5	6	7
	3	7	8	9	10

7. Beam Element

The y''^2 appearing in the total potential energy expression (3.56) for the thin elastic beam necessitates a C^1 trial function for it. Such an admissible trial function is provided by the piecewise cubic in Eqs. (4.45) and (4.46). These shape functions are used in

$$(k_e)_{ij} = \int_{x_1}^{x_1+h} p(x)\phi_i''\phi_j'' \, dx \tag{4.51}$$

or

$$(k_e)_{ij} = \frac{1}{h^3} \int_0^1 p(\xi)\phi_i''\phi_j'' \, d\xi \tag{4.52}$$

to form, with $p(x) = 1$, the beam element stiffness matrix

$$k_e = \frac{1}{h^3} \begin{bmatrix} 12 & 6h & -12 & 6h \\ 6h & 4h^2 & -6h & 2h^2 \\ -12 & -6h & 12 & -6h \\ 6h & 2h^2 & -6h & 4h^2 \end{bmatrix} \tag{4.53}$$

which is relative to the element nodal value arrangement (y_1, y_1', y_2, y_2'), $y' = dy/dx$. As with the string element, here also h can be removed from the inside of k_e with $\theta = hy'$.

Notice that the rank of k_e in Eq. (4.53) is only 2 since $y = 1$ and $y = x$, the *rigid body modes* of the beam, are not storing elastic energy in it.

8. Complex Structures

The popularity of the finite element method among structural engineers is much due to its ability to routinely deal with complicated elastic structures. A variational, or energetic, formulation is adopted to describe the problem and the total potential energy is computed by the addition of the elastic energies of all structural members. As an example of this consider the two dimensional truss in Fig. 4.11, comprising seven pin-joined elastic bars (elements). Each joint or node has the freedom to move only in the plane and its displacement is fixed by the displacement pair (u, v) in the x and y directions, respectively. In order to compute the elastic energy in a typical element,

Fig. 4.11 Seven bar truss.

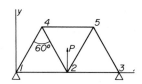

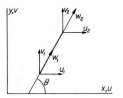

Fig. 4.12   Truss member in the xy plane.

the element is isolated as in Fig. 4.12 and the elastic energy in it is computed. Elastic energy is stored in the elastic bar only by axial stretching and is equal to

$$E = \tfrac{1}{2}p(w_2 - w_1)^2 = \tfrac{1}{2}w_e{}^T k_e w_e \tag{4.54}$$

where

$$k_e = p \begin{bmatrix} 1 & -1 \\ -1 & 1 \end{bmatrix}, \; w_e = \begin{bmatrix} w_1 \\ w_2 \end{bmatrix} \tag{4.55}$$

and where p is a constant involving the geometrical and elastic properties of the bar. To express the element matrix in terms of the displacements (u, v) in the direction of (x, y), common to all elements, we observe that

$$w_1 = u_1 \cos \theta + v_1 \sin \theta$$
$$w_2 = u_2 \cos \theta + v_2 \sin \theta \tag{4.56}$$

Or in matrix form $w_e = T y_e$

$$\begin{bmatrix} w_1 \\ w_2 \end{bmatrix} = \begin{bmatrix} c & s & & \\ & & c & s \end{bmatrix} \begin{bmatrix} u_1 \\ v_1 \\ u_2 \\ v_2 \end{bmatrix} \tag{4.57}$$

where $c = \cos \theta$ and $s = \sin \theta$. Now

$$w_e{}^T k_e w_e = y_e{}^T (T^T k_e T) y_e \tag{4.58}$$

and the element stiffness matrix relative to $y_e{}^T = (u_1, v_1, u_2, v_2)$ becomes

$$k_e = p \begin{bmatrix} c^2 & cs & -c^2 & -cs \\ cs & s^2 & -cs & -s^2 \\ -c^2 & -cs & c^2 & cs \\ -cs & -s^2 & cs & s^2 \end{bmatrix} \tag{4.59}$$

The rank of this element matrix is only 1 since rigid body displacements parallel to the x axis, y axis, and rotation are not energy storing, and k_e in

Eq. (4.59) can also be written as

$$k_e = p(c, s, -c, -s)^{\mathrm{T}}(c, s, -c, -s) \qquad (4.60)$$

As soon as the general element stiffness matrix is available for a typical element, the assembly of all seven elements into the global stiffness matrix K is routinely carried out from the local–global numbering correspondance table

	local	
	1	2
1	1	2
2	2	3
3	1	4
element 4	4	2
5	2	5
6	3	5
7	4	5

EXERCISES

1. Develop the element stiffness and mass matrices k_e and m_e, as well as the element load vector f_e for $-y'' + y = 1$, for the cubic four nodal point element of Fig. 4.13.

    Fig. 4.13 Cubic four nodal point string element. ⌊ *h* 2 *h* 3 *h* 4 ⌋

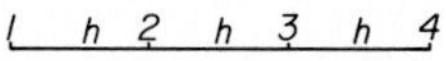

2. Generate the global stiffness and mass matrices K and M, as well as the load vector f, from two cubic string finite elements of Exercise 1.

3. Introduce the essential boundary condition $y(0) = 0$ into the linear system you obtained in Exercise 2 and solve the system on the computer.

4. Form the element stiffness and mass matrices for a circular membrane element. Consult Exercise 1, Chapter 3, for this.

5. Use two finite elements to solve the beam problem of Fig. 4.14.

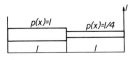

Fig. 4.14   Sudden change of cross section in a beam.

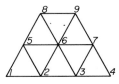

Fig. 4.15   Sixteen bar truss.

6.   For the truss in Fig. 4.15 write down the correspondance between the local and global numbers of the 16 elements.

7.   A variational formulation to a boundary value problem is readily set up with *least squares*. As an illustration to this technique, consider the equation $-y'' + y = 1$, $y(0) = 0$, $y(1) + y'(1) = 0$. Reduce the order of this equation by writing

$$y' + y = z$$
$$z' - z = -1$$

and associate with this pair of equations the error integral

$$\pi(y, z) = \int_0^1 \left[(y' + y - z)^2 + (z' - z + 1)^2 \right] dx$$

Use first-order finite elements to approximate *both* y and z and minimize (y, z). Compare your result with that obtained from the principle of minimum potential energy. (Least squares preceded Ritz and can be traced back to J. Boussinesq's work on the theory of heat in 1792.)

SUGGESTED FURTHER READING

Zienkiewicz, O. C., *The Finite Element Method in Engineering Science*. McGraw–Hill, New York, 1971.

Desai, C. S., and Abel, J. F., *Introduction to the Finite Element Method*. Van Nostrand Reinhold, Princeton, New Jersey, 1972.

Norrie, D. H., and de Vries, G., *The Finite Element Method*. Academic Press, New York, 1973.

Pilkey, W., Saczalski, K., and Schaeffer, H., *Structural Mechanics Computer Programs*. Univ. of Virginia Press, Charlottesville, 1974.

Cook, R. D., *Concepts and Applications of Finite Element Analysis*. Wiley, New York, 1974.

Gallagher, R. H., *Finite Element Analysis*. Prentice–Hall, Englewood Cliffs, New Jersey, 1975.

Heubner, K. H., *The Finite Element Method for Engineers*. E. Wiley, New York, 1975.

5 _Discretization Accuracy_

1. Energy Theorems

Some new notations and concepts are in order at this point. First to be introduced is the *inner* or *scalar product*

$$(y, z) = \int_0^1 y(x)z(x)\, dx \tag{5.1}$$

between the two functions $y(x)$ and $z(x)$, where $0 \leqslant x \leqslant 1$, that have a finite $\|\cdot\|_0$ norm. Obviously

$$(y, y) = \|y\|_0^2 \tag{5.2}$$

Next, Eq. (5.1) is generalized to include the *energy inner product* $E(y, z)$

$$E(y, z) = \int_0^1 [p(x)y'z' + q(x)yz]\, dx \tag{5.3}$$

between the two functions $y(x)$ and $z(x)$, where $0 \leqslant x \leqslant 1$, that have a finite $\|\cdot\|_1$ energy norm. When $y = z$

$$E(y, y) = \int_0^1 [p(x)y'^2 + q(x)y^2]\, dx \tag{5.4}$$

and

$$E(y, y) = \|y\|_1^2 \tag{5.5}$$

A generalization of $E(y, z)$ to the elastic beam energy is obvious.

To prove the *Cauchy–Schwarz inequality* in the energy product, observe that

$$E(y + \alpha z, y + \alpha z) = E(y, y) + 2\alpha E(y, z) + \alpha^2 E(z, z) \geqslant 0 \tag{5.6}$$

for any value of the scalar α. Equation (5.6) is a nonnegative quadratic in α, its discriminant is necessarily not positive, and

$$E^2(y, z) - E(y, y)E(z, z) \leqslant 0 \tag{5.7}$$

Consequently,

$$|E(y, z)| \leqslant \|y\|_m \|z\|_m \tag{5.8}$$

which is the celebrated inequality.

Setting $\alpha = 1$ in Eq. (5.6) reduces it to the form

$$\|y + z\|_m^2 = \|y\|_m^2 + 2E(y, z) + \|z\|_m^2 \tag{5.9}$$

or

$$\|y + z\|_m^2 \leqslant \|y\|_m^2 + 2|E(y, z)| + \|z\|_m^2 \tag{5.10}$$

which, with the Cauchy–Schwarz inequality (5.8), becomes the *triangle inequality*

$$\|y + z\|_m \leqslant \|y\|_m + \|z\|_m \tag{5.11}$$

and from which another useful inequality

$$\|y - z\|_m \geqslant \big| \|y\|_m - \|z\|_m \big| \tag{5.12}$$

follows.

Some simple, though formal, proofs can be given to the following useful theorems using the triangle inequality (5.11):

(i)   *If* $\lim_{i \to \infty} \|y - y_i\| = 0$ *and* $\lim_{j \to \infty} \|y - y_j\| = 0$, *then also* $\lim_{i, j \to \infty}$ $\|y_i - y_j\| = 0$. Indeed $\|y_i - y_j\| \leqslant \|y - y_i\| + \|y - y_j\|$.

(ii)   *Convergence in* $\|\cdot\|_m$ *cannot be to two different limits.* Indeed if $\lim_{i \to \infty} \|y - y_i\|_m = 0$ and $\lim_{i \to \infty} \|z - y_i\|_m = 0$, then since $\|y - z\|_m \leqslant \|y - y_i\|_m + \|z - y_i\|_m$, it follows that $y = z$.

(iii)   *Convergence in the energy implies convergence of the energies; that is,* if $\lim_{i \to \infty} \|y - y_i\|_m = 0$, then also $\lim_{i \to \infty} \|y_i\|_m = \|y\|_m$. This follows directly from Eq. (5.12).

In terms of the energy inner product $E(y, z)$, the total potential energy may be written as

$$\pi(y) = \tfrac{1}{2}E(y, y) - (f, y) \tag{5.13}$$

In this notation, the vanishing of the first variation $\delta\pi(y)$ of the total potential $\pi(y)$, taken with respect to the energetically admissible z (i.e., satisfying the essential boundary conditions and with $E(z, z) < \infty$), leads to

$$E(y, z) = (f, z) \tag{5.14}$$

which is the equation of *virtual displacements*, z being that displacement. The same holds for the finite element (Ritz) trial function $\tilde{y}$ and the optimal $\hat{y}$ that minimizes $\pi(\tilde{y})$. For this pair of functions

$$E(\tilde{y}, \hat{y}) = (f, \tilde{y}) \tag{5.15}$$

and

$$\pi(\hat{y}) = -\tfrac{1}{2}E(\hat{y}, \hat{y}) = -\tfrac{1}{2}(f, \hat{y}) \tag{5.16}$$

It is also true for the solution y of the boundary value problem with which $\pi(y)$ is associated that

$$\pi(y) = -\tfrac{1}{2}E(y, y) = -\tfrac{1}{2}(f, y) \tag{5.17}$$

If the load function $f(x)$ is a unit point force (impulse, unit delta function) at $x = x_0$, then according to Eqs. (5.16) and (5.17)

$$\pi(y) = -\tfrac{1}{2}y(x_0), \qquad \pi(\hat{y}) = -\tfrac{1}{2}\hat{y}(x_0) \tag{5.18}$$

Assuming $y(x_0)$ is finite, the principle of minimum potential energy implies that

$$y(x_0) \geqslant \hat{y}(x_0) \tag{5.19}$$

Because $E(y, y) > 0$, it ensues from Eqs. (5.16) and (5.17) that $y(x_0)$ and $\hat{y}(x_0)$ are positive, and the magnitude of $y(x_0)$ is never less than that of $\hat{y}(x_0)$ when $f(x)$ is a point force at $x = x_0$.

Let $\tilde{y}$ be the finite element *trial space* from which the optimal $\hat{y}$ is selected through the minimization of $\pi(\tilde{y})$. Substituting $z = \tilde{y}$ in Eq. (5.14) and subtracting it from Eq. (5.15) produces

$$E(y - \hat{y}, \tilde{y}) = 0 \tag{5.20}$$

and the error function $\hat{e} = y - \hat{y}$ in the finite element solution is *orthogonal in the energy* to any member of the space of trial functions $\tilde{y}$.

From the Cauchy–Schwarz inequality we have that

$$|E(\hat{e}, z)| \leqslant \|\hat{e}\|_m \|z\|_m \tag{5.21}$$

where $\hat{e} = y - \hat{y}$. Let z satisfy the string or beam differential equations $-(p(x)z')' + q(x)z = f(x)$ or $(p(x)z'')'' + q(x)z = f(x)$ with their boundary conditions such that

$$E(z, \hat{e}) = (f, \hat{e}) \tag{5.22}$$

is the equation of virtual displacements of these boundary value problems. From Eqs. (5.22) and (5.21) we have that

$$|(f, \hat{e})| \leqslant \|\hat{e}\|_m \|z\|_m \tag{5.23}$$

which, by choosing $f(x) = \hat{e}(x)$, becomes

$$\|\hat{e}\|_0^2 \leqslant \|\hat{e}\|_m \|z\|_m \tag{5.24}$$

an inequality which is helpful in relating $\|\hat{e}\|_0$ to $\|\hat{e}\|_m$. If we choose for $f(x)$ in Eq. (5.23) a unit delta function at point $x = x_0$, it becomes

$$|y(x_0) - \hat{y}(x_0)| \leqslant \|\hat{e}\|_m \|z\|_m \tag{5.25}$$

which is helpful in relating the *pointwise* error $|y - \hat{y}|$ to the energy error $\|y - \hat{y}\|_m$, where $m = 1$ for the string and $m = 2$ for the beam.

2.   Energy Rates of Convergence

Having discussed in Chapter 4 the technical or procedural aspects of finite elements, we now turn to the basic theoretical question of the method concerning the convergence of its solution to the analytical one. Because the finite element solution $\hat{y}$ minimizes the energy error $\|y - \tilde{y}\|_m$, this norm presents itself as the most natural candidate for the finite element error measurement, and from which the error in other norms will have to be deduced.

Our aim is to express $\|y - \hat{y}\|_m$ in terms of the intrinsic and discretization parameters and properties of the problem. To this end, we make use of the basic inequality

$$\|y - \hat{y}\|_m \leqslant \|y - \tilde{y}\|_m \tag{5.26}$$

and estimate $\|y - \tilde{y}\|_m$ with a reasonable, though convenient, trial function $\tilde{y}$, obtaining in this way an upper bound on the error $\|y - \hat{y}\|_m$ in the finite element solution. A sensible choice for $\tilde{y}$ in Eq. (5.26) is the hypothetical *interpolate* to the analytic solution y such that *at the nodal points* $y = \tilde{y}$. As the finite element mesh duplicates itself over the elements, it is enough that we examine only one typical element, estimate the energy error for it, and sum up these errors over all the Ne finite elements in the mesh.

First we shall consider in detail the accuracy of the first-order string element in Fig. 4.1, and look at the isolated element of size h shown in Fig. 5.1. We adopt the notation $\tilde{e}(x) = y(x) - \tilde{y}(x)$, $\hat{e}(x) = y(x) - \hat{y}(x)$, $\tilde{e}'(x) = y'(x) - \tilde{y}'(x)$, $\hat{e}'(x) = y'(x) - \hat{y}'(x)$, and have for the *interpolating* $\tilde{y}$ that $\tilde{e}(x_1) = \tilde{e}(x_2) = 0$ at the nodal points. In our present analysis we assume that y is analytic inside each element or that all derivatives of y are finite in the *interior* of the elements. Discontinuities are naturally placed at the inter element nodal points.

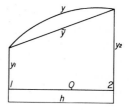

Fig. 5.1 Error distribution $y - \bar{y}$ in an interpolating linear element.

By *Rolle's* theorem, there is a point x_0 inside the typical element at which $\bar{e}'(x) = 0$, and at which $|\bar{e}(x)|$ is maximal. Expansion of $\bar{e}(x)$ around this point x_0 in a Taylor series is, up to $(x - x_0)^2$ is of the form

$$\bar{e} = \bar{e}_0 + (x - x_0)\bar{e}_0' + \tfrac{1}{2}(x - x_0)^2 y''(\xi) \tag{5.27}$$

where ξ is somewhere between x_0 and x. Since $\bar{y}$ is linear inside the element, $\bar{e}'' = y''$ and hence the y'' in Eq. (5.27). By letting x in Eq. (5.27) be x_1 or x_2, whichever is nearer to x_0, we have that $\bar{e}(x) = 0$, where $|x - x_0| \leqslant \tfrac{1}{2}h$, and consequently

$$|\bar{e}_0| \leqslant \tfrac{1}{8}h^2 |y''(\xi)| \tag{5.28}$$

or

$$\max|\bar{e}| \leqslant \tfrac{1}{8}h^2 \max|y''| \tag{5.29}$$

where $\max| \,|$ refers to the maximal value *inside* the element. To estimate $|\bar{e}'|$ inside the element, we expand $\bar{e}'$ around $x = x_0$, at which $\bar{e}' = 0$, and have that

$$\bar{e}' = (x - x_0)y''(\xi) \tag{5.30}$$

Next we choose x as the point where $|\bar{e}'|$ is maximal and get

$$\max|\bar{e}'| \leqslant h \max|y''| \tag{5.31}$$

Introduction of Eqs. (5.29) and (5.31) into

$$\|y - \hat{y}\|_1^2 \leqslant \max|y' - \bar{y}'|^2 \int_0^1 p(x)\,dx + \max|y - \bar{y}|^2 \int_0^1 q(x)\,dx \tag{5.32}$$

produces from Eqs. (5.4) and (5.26) the desired error estimate

$$\|y - \hat{y}\|_1^2 \leqslant (c_1 h^2 + c_2 h^4)\max|y''|^2 \tag{5.33}$$

where c_1 and c_2 are two constants independent of y and h and $\max|y''|$ is over $0 \leqslant x \leqslant 1$, but only in the interior of the elements. In the event that y'' is finite within all elements, Eq. (5.33) predicts that $\|y - \hat{y}\|_1 \leqslant O(h)$, and the energy convergence of the finite element method with first-order elements is proved; as $h \to 0$ so inevitably does $\|y - \hat{y}\|_1$.

Now we shall analyse the accuracy of the quadratic finite element in a manner that will point to the generalization of the error estimate for elements of any order. Here $\bar{y} = y_1\phi_1 + y_2\phi_2 + y_3\phi_3$, with the three shape functions ϕ_1, ϕ_2, and ϕ_3 given in Eq. (4.41). Once more, $\bar{y}$ is taken to be the interpolate to y with exact y_1, y_2, and y_3. Expansion of y_1 and y_3 around the central node number 2 results in

$$\bar{y} = y_2(\phi_1 + \phi_2 + \phi_3) + hy_2'(\phi_3 - \phi_1) + \tfrac{1}{2}h^2y_2''(\phi_1 + \phi_3) + O(y'''h^3) \quad (5.34)$$

But

$$\phi_1 + \phi_2 + \phi_3 = 1, \qquad \phi_3 - \phi_1 = \xi, \qquad \phi_1 + \phi_3 = \xi^2 \qquad (5.35)$$

and therefore, since $x = h\xi$

$$\bar{y} = y_2 + xy_2' + \tfrac{1}{2}x^2y_2'' + O(y'''h^3) \qquad (5.36)$$

Also

$$y = y_2 + xy_2' + \tfrac{1}{2}x^2y_2'' + O(y'''x^3) \qquad (5.37)$$

Hence, by subtracting Eqs. (5.36) and (5.37), we have that

$$\max|y - \bar{y}| \leqslant ch^3 \max|y'''| \qquad (5.38)$$

Similarly,

$$\max|y' - \bar{y}'| \leqslant ch^2 \max|y'''| \qquad (5.39)$$

where in Eqs. (5.38) and (5.39) c is a generic constant independent of h and y. Consequently, for the quadratic element

$$\|y - \hat{y}\|_1^2 \leqslant (c_1h^4 + c_2h^6)\max|y'''|^2 \qquad (5.40)$$

where the maximum of y''' is over $0 \leqslant x \leqslant 1$ inside the elements only. When y''' is finite in the interior of the elements, $\|y - \hat{y}\|_1 \leqslant O(h^2)$ and doubling of the *convergence rate* is expected for quadratic elements over that of linear elements. If $\|y - \hat{y}\|_1$ actually behaves as the estimated bounds do, then with linear elements by doubling the number of elements the error in the energy norm declines by a factor of 2, while with quadratic elements the same error declines by a factor of 4.

The fortunate choice of polynomials in the shape functions becomes more striking in light of the present error analysis. Not only is it easier to differentiate and integrate polynomials, but they are also most appropriate for the approximation. According to Taylor's theorem any analytic function is polynomiallike over a small interval. In case the interpolate $\bar{y}$ is, say, quadratic inside the element and y smooth or analytic, their Taylor expansions around a point inside the element include the same first quadratic terms, and $|y - \bar{y}| = O(h^3y''')$. One power of h is lost in differentiation and $|y' - \bar{y}'| =$

$O(h^2 y''')$. In the more general situation in which the shape functions include all the terms of a (complete) polynomial of degree p

$$\max|y - \tilde{y}| \leqslant O[h^{p+1} y^{(p+1)}]$$

$$\max|y' - \tilde{y}'| \leqslant O[h^p y^{(p+1)}] \qquad (5.41)$$

$$\max|y^{(m)} - \tilde{y}^{(m)}| \leqslant O[h^{p+1-m} y^{(p+1)}]$$

and in general,

$$\|y - \tilde{y}\|_m \leqslant ch^{p+1-m} \max|y^{(p+1)}| \qquad (5.42)$$

where the maximum of $y^{(p+1)}$ is sought over the entire x interval, but only in the interior of the elements. In the next section we shall show that the error bound in Eq. (5.42) is optimal in its dependence upon h in the sense that equality may actually occur in it.

Equation (5.42) not only predicts a higher rate of convergence with higher-order elements, it also indicates that the best mesh grading is such that $h^{p+1-m}|y^{(p+1)}|$ is constant over the elements. Also, since $\max|y^{(p+1)}|$ is searched inside the elements only, any discontinuity in y due to the discontinuity of $p(x)$, $q(x)$, or $f(x)$ should be confined to the interelement nodal points.

As an example of the error analysis of this section, and its conclusions, consider a circular plate ($m = 2$) discretized with cubic ($p = 3$) C^1 finite elements. Equation (5.42) predicts here that $\|y - \hat{y}\|_2 \leqslant O(h^2 y'''')$ provided y'''' is finite. Let the plate be point loaded at the center. This singular load causes y to become singular at $r = 0$ to the extent that $y = r^2 \log r$ near the origin, and y barely possesses two finite derivatives close to $r = 0$, barely since $\log r$ is an extremely weak singularity. Hence, with a *uniform* mesh of Ne finite elements, each of size $h = 1/Ne$, we expect no more than $\|\hat{e}\|_2^2 = O(h^2) = O(Ne^{-2})$ instead of the full rate $\|\hat{e}\|_2^2 = O(h^4)$ when y'''' is finite. According to the analysis of Section 1, Eq. (5.20) with a point load,

$$\|y - \hat{y}\|_2^2 = E(y - \hat{y}, y - \hat{y}) = E(y - \hat{y}, y) = E(y, y) - E(y, \hat{y})$$

$$= (f, y) - (f, \hat{y}) = (f, y - \hat{y}) = y(0) - \hat{y}(0) \qquad (5.43)$$

and the energy convergence may be conveniently measured by $y(0) - \hat{y}(0)$. Figure 5.2a shows the numerical results for a clamped plate point loaded at the center and discretized with a uniform mesh of cubic finite elements, and indeed the error $\delta y_1 = y(0) - \hat{y}(0)$, which is proportional to the energy error, is only $O(h^2) = O(Ne^{-2})$. Regarding nonuniform meshes, when the mesh is graded by the rule $h_e = he^{3/2}$, with the smallest element near the singular origin, the full rate of convergence $\|e\|_2^2 = O(Ne^{-4})$ is nearly recovered, as shown in Fig. 5.2b. Figure 5.3 depicts the pointwise error $\delta y(r) = y(r) - \hat{y}(r)$

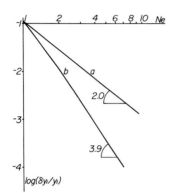

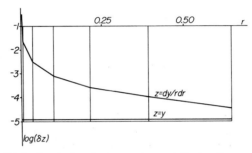

Fig. 5.2   Clamped circular plate point loaded at the center and discretized with cubic elements. Curve (a) shows the convergence of the central deflection y_1 for a uniform mesh, and curve (b) for a mesh graded by the rule $h_e = he^{3/2}$, with the smallest element near the origin.

Fig. 5.3   Error distribution in the displacement y and the bending moment $dy/r\,dr$ in the circular plate of Fig. 5.2(b).

in the deflection and the error $(y' - \hat{y}')/r$ in the bending moment. It is interesting that the error in the deflection is practically constant along r, including $r = 0$, while that in the bending moment is mostly confined to the immediate neighborhood of the singularity.

In practice, the variation of $y^{(p+1)}$ along x is only roughly known beforehand if at all. One proceeds by putting down a mesh graded, as well as one can, refining and grading it according to the behavior of the *computed* solution. In theory at least, it is possible to minimize $\pi(\hat{y})$ also with respect to the nodal point locations, but the resulting algebraic system is nonlinear, complicated to solve, and this approach to mesh grading is of dubious practical value

It should be borne in mind that Eq. (5.42) does not forecast a *monotonic* energy convergence as the number of elements is increased since the energy error depends not only on the number of elements but also on the location of the nodes. But if the previous mesh is included in the finer, as when some elements are halved without changing the location of the previous nodes, then the total potential energy of the finer mesh cannot increase and convergence is monotonic.

3. Sharpness of the Energy Error Estimate

The basic energy error estimate of the last section was obtained from the particular $\tilde{y}$ chosen to be the interpolate of y, which led us to the error estimate $\|y - \hat{y}\|_m \leqslant O(h^{p+1-m})$. It is left to prove that, as far as the power of h, or the *rate* of convergence, is concerned, this is the optimal error estimate. There is no other finite element approximation in the space $\tilde{y}$ giving a *higher* power to h than $p + 1 - m$. Our theoretical prediction of the finite element rate of convergence is therefore sharp. We do this proof by choosing some smooth function y (i.e., one that has as many derivatives as needed in the error estimate), and minimizing $\|y - \tilde{y}\|_m$ *over only one element without imposing any continuity restrictions at its ends*. The approximation obtained in this way is the *best energy fit* for the element, constituting a lower bound on the actual error in the energy.

For the linear string element, we choose y to be the next-order polynomial—a quadratic of the form $y = \frac{1}{2}\alpha x^2$—and minimize

$$\|y - \tilde{y}\|_1^2 = \int_0^1 (y' - \tilde{y}')^2 \, dx \tag{5.44}$$

with respect to a_0 and a_1 in $\tilde{y} = a_0 + a_1 x$. With these y and $\tilde{y}$, Eq. (5.44) becomes

$$\|y - \tilde{y}\|_1^2 = \int_0^h (\alpha x - a_1)^2 \, dx \tag{5.45}$$

and the best energy fit y ensues from $\partial\|y - \tilde{y}\|_1/\partial a_1 = 0$, resulting in $a_1 = \frac{1}{2}\alpha h$ and, for one element,

$$\|y - \hat{y}\|_1^2 = \frac{1}{12}\alpha^2 h^3 \tag{5.46}$$

For $Ne = 1/h$ elements, $\|\hat{e}\|_1 = O(h)$, as theoretically predicted by Eq. (5.42).

A thin elastic beam is associated with

$$\|y - \tilde{y}\|_2^2 = \int_0^1 (y'' - \tilde{y}'')^2 \, dx \tag{5.47}$$

and we choose to minimize it when cubic elements are involved with $y = \frac{1}{24}\alpha x^4$ and $\tilde{y} = a_0 + a_1 x + a_2 x^2 + a_3 x^3$, and have for one element that

$$\|y - \tilde{y}\|_2^2 = \int_0^h \left(\frac{1}{2}\alpha x^2 - 2a_2 - 6a_3 x\right)^2 dx \tag{5.48}$$

Minimization of $\|y - \tilde{y}\|_2^2$ with respect to a_2 and a_3 delivers the values $a_2 = -\frac{1}{24}\alpha h^2$ and $a_3 = \frac{1}{2}\alpha h$ and for a single element,

$$\|y - \hat{y}\|_2^2 = \frac{1}{720}\alpha h^5 \tag{5.49}$$

For the complete mesh with $Ne = 1/h$ finite elements, $\|y - \hat{y}\|_1^2 = O(h^4)$, as correctly predicted in Eq. (5.42).

An interesting corollary to the above discussion is that when $-y'' = f(x)$ and $y'''' = f(x)$, with boundary conditions that add no boundary terms to $\pi(y)$, are solved with linear and cubic finite elements, respectively, the nodal values *computed with these finite elements are exact*, provided that all integrations in $\pi(y)$ are accurate. To prove this, let $\tilde{y} = a_0 + a_1 x$ inside the linear element so that the best energy fit is selected by minimizing

$$\|y - \tilde{y}\|_1^2 = \int_{x_1}^{x_1 + h} (y' - a_1)^2 \, dx \tag{5.50}$$

resulting in

$$a_1 = h^{-1}[y(x_1 + h) - y(x_1)] \tag{5.51}$$

which means that the best fit of the linear segment occurs *when it is parallel to the chord*. The trial function $\tilde{y}$, consisting of all the chord segments, that is, the interpolate, is energetically admissible and satisfies the essential boundary conditions. Since each segment is the best fit, the assembly of all segments minimizes the total potential energy and constitutes the finite element solution. The interpolate is then the solution and is, of course, exact at the nodes.

Similarly, for the cubic beam element, for which

$$\|y - \tilde{y}\|_2^2 = \int_{x_1}^{x_2} (y'' - 2a_2 - 6a_3 x)^2 \, dx \tag{5.52}$$

$\|\tilde{e}\|_2^2$ is minimized with respect to a_2 and a_3, yielding

$$\int_{x_1}^{x_2} (y'' - 2a_2 - 6a_3 x) \, dx = 0, \qquad \int_{x_1}^{x_2} (y'' - 2a_2 - 6a_3 x) x \, dx = 0 \tag{5.53}$$

which become, since $y''x = (xy')' - y'$,

$$y'(x_2) - y'(x_1) = 2a_2(x_2 - x_1) + 3a_3(x_2^2 - x_1^2) \tag{5.54}$$

and

$$[x_2 y'(x_2) - x_1 y'(x_1)] - [y(x_2) - y(x_1)] = a_2(x_2^2 - x_1^2) + 2a_3(x_2^3 - x_1^3) \tag{5.55}$$

At the nodes $x = x_1$ and $x = x_2$, we have from $\tilde{y} = a_0 + a_1 x + a_2 x^2 + a_3 x^3$ that when $\tilde{y}$ interpolates y

$$\begin{aligned} y'(x_1) &= a_1 + 2a_2 x_1 + 3a_3 x_1^2 \\ y'(x_2) &= a_1 + 2a_2 x_2 + 3a_3 x_2^2 \end{aligned} \tag{5.56}$$

which are seen to solve Eqs. (5.54) and (5.55). We conclude that the cubic $\tilde{y}$ in the beam element, used to discretize $y'''' = f(x)$, provides exact y and y' values at the nodes in the finite element solution.

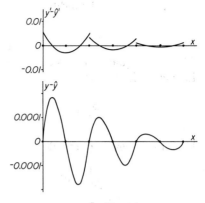

Fig. 5.4 Error distribution in y and y' for $-y'' + y = 1$, $y(0) = 0$, $y'(1) = 0$ discretized with three quadratic finite elements. The superconvergence points for y are at the nodes while for y' they are the Gauss points.

This phenomenon is referred to as *superconvergence*, and the points at which the accuracy is higher are the superconvergence points. For $-y'' = f(x)$ and $y'''' = f(x)$, discretized with linear and cubic elements, the super-convergence points for y and, y and y', respectively, are the nodes. Figure 5.4 shows the error distribution of $\hat{e} = y - \hat{y}$ and $\hat{e}' = y' - \hat{y}'$ for $-y'' + y = 1$, $y(0) = 0$, and $y'(1) = 0$ discretized with three quadratic finite elements. In this figure the nodes are seen to be the superconvergence points for y, while the *Gauss points* $\xi = \pm\sqrt{3}/3$, in an element that extends over $-1 \leqslant \xi \leqslant 1$, are seen to be the superconvergence points for y'. The location of the superconvergence points may be predicted as follows: assume $y = \alpha\xi^3$ and $\tilde{y} = y_1\phi_1 + y_2\phi_2 + y_3\phi_3$, $\phi_1 = \xi(\xi - 1)/2$, $\phi_2 = 1 - \xi^2$, $\phi_3 = \xi(\xi + 1)/2$, $-1 \leqslant \xi \leqslant 1$. If the best energy fit $\hat{y}$ to y is the interpolate, then $y' = \hat{y}'$ at $\xi = \pm\sqrt{3}/3$. It can further be shown that the Gauss points are special points also in higher-order elements.

4. L_2 Error Estimate

After having established the theoretical rate of convergence of the finite element method in the $\|\cdot\|_m$ energy norm, we turn to the subsequent problem of relating this error estimate to the L_2 and eventually to the L_∞, or point-wise, error estimates.

As a first step in this direction, we prove that the *differential operators* we previously encountered in the boundary value problems are *positive bounded from below* meaning that

$$\|y\|_0 \leqslant c\|y\|_m, \qquad c > 0 \tag{5.57}$$

for all energetically admissible functions y [i.e., those that may be used in $\pi(y)$]. To prove Eq. (5.57) for the $\|y\|_1$ norm, for the particular situation in which the essential conditions are only $y(0) = 0$ we write

$$y(x) = \int_0^x y'(z)\,dz \tag{5.58}$$

Then we square both sides of Eq. (5.58) and use the Cauchy–Schwarz inequality in the form

$$y^2(x) = \left| \int_0^x 1 \cdot y'(z)\,dz \right|^2 \leqslant \int_0^x 1^2\,dz \int_0^x y'^2\,dz \tag{5.59}$$

Next we integrate both sides of the above equation and get

$$\int_0^1 y^2\,dx \leqslant \int_0^1 y'^2\,dx \int_0^1 x\,dx = \tfrac{1}{2} \int_0^1 y'^2\,dx \tag{5.60}$$

Furthermore, since

$$\|y\|_1^2 \geqslant p_{\min} \int_0^1 y'^2\,dx, \qquad p_{\min} = \min_x \{p(x)\} \tag{5.61}$$

it ensues, if $p_{\min} > 0$, that

$$\|y\|_0^2 \leqslant \tfrac{1}{2} p_{\min}^{-1} \|y\|_1^2 \tag{5.62}$$

which is our desired result.

Equation (5.62) provides the proof of the positive definiteness of the global stiffness matrix K set up by the Ritz method, even when $q(x) = 0$. In fact, according to Eq. (5.62),

$$a^{\mathrm{T}} K a \geqslant 2 a^{\mathrm{T}} M a p_{\min} \tag{5.63}$$

and since M obtained from Eq. (4.12) with $q = 1$ is positive definite by virtue of the linear independence of the basis functions so must K be. Positive definiteness is another intrinsic property of the boundary value problems we consider that is preserved in the discretization.

Concerning the convergence of the finite element solution *in the mean*, or in the $\|\cdot\|_0$ norm, Eq. (5.57) indicates that in the second-order problem $\|\hat{e}\|_0 \leqslant c\|\hat{e}\|_1$ since $\hat{e}(x)$ is an energetically admissible trial function, and hence with polynomial elements of order p, $\|\hat{e}\|_0 \leqslant O(h^p)$. But we are skeptical about the sharpness of this error estimate. Because $\|\cdot\|_0$ includes no differentiations that use powers of h, we expect $\|\hat{e}\|_0$ to be $O(h^{p+1})$. To obtain a better $\|\hat{e}\|_0$ error estimate, we avail ourselves of Eq. (5.24) in assuming z to be the solution of $-(p(x)z')' + q(x)z = \hat{e}(x)$. Corresponding to this equation, the equation of virtual displacements may be written as $E(z, \hat{e}) = \|\hat{e}\|_0^2$. From Eq. (5.20) we have that $E(\tilde{y}, \hat{e}) = 0$, and hence the subtraction of the last two equations produce $E(z - \tilde{y}, \hat{e}) = \|\hat{e}\|_0^2$, from which we have by the Cauchy–Schwarz inequality that $\|\hat{e}\|_0^2 \leqslant \|\hat{e}\|_1\|z - \tilde{y}\|_1$. At this point it is shown, not by us, that since $\hat{e}(x)$ is the load, a $\tilde{y}$ can be found such that

$\|z - \tilde{y}\|_1 \leqslant ch\|\hat{e}\|_0$. We are left with $\|\hat{e}\|_0 \leqslant ch\|\hat{e}\|_1$ that with linear elements results in $\|\hat{e}\|_0 = O(h^2)$ rather than $\|\hat{e}\|_0 = O(h)$ from Eq. (5.62).

Successive applications of Eq. (5.60) produce

$$\int_0^1 y^2 \, dx \leqslant \tfrac{1}{2} \int_0^1 y'^2 \, dx \leqslant \tfrac{1}{4} \int_0^1 y''^2 \, dx \tag{5.64}$$

from which

$$\|y\|_0^2 \leqslant \tfrac{1}{4} p_{\min}^{-1} \|y\|_2^2 \tag{5.65}$$

emerges for the fourth-order beam problem. It is concluded from Eq. (5.65) that

$$\|\hat{e}\|_0 \leqslant c\|\hat{e}\|_2 \tag{5.66}$$

and the finite element solution to the beam problem that converges energetically converges also in the mean. The positiveness from below of the differential operator of the beam problem as expressed in Eq. (5.65), guarantees the positive definiteness of the global stiffness matrix assembled from beam elements.

5. L_∞ Error Estimate

To relate the pointwise error to the energy error, assume in Eq. (5.59) that x is the point at which $y^2(x)$ is maximal. Then

$$\max_x \{y^2(x)\} \leqslant \int_0^1 y'^2 \, dx \tag{5.67}$$

or

$$\|y\|_\infty \leqslant p_{\min}^{-1/2} \|y\|_1 \tag{5.68}$$

and as a result $\|\hat{e}\|_\infty \leqslant O(h^p)$. A similar analysis can be carried out for the beam.

Another way to prove the pointwise convergence of the finite element solution is via Eq. (5.23). In case the response z to a point force at x_0 is of finite energy, which is not the case, for example, for a circular membrane, Eq. (5.25) is valid, $\|z\|_m$ is independent of h, and $|y(x_0) - \hat{y}(x_0)| \leqslant c\|\hat{e}\|_m$. Hence, if x_0 is the point where $|\hat{e}(x_0)|$ is maximum, $\|\hat{e}\|_\infty \leqslant O(h^{p+1-m})$. If not optimal, this proves at least that the pointwise rate of convergence is no less than the rate of convergence in the energy norm.

A better estimate for $|\hat{e}|$ can be obtained sometimes as follows: Let y be the solution to our analytic boundary value problem, $\tilde{y}$ the relevant finite element trial space, and $\hat{y}$ the energetically best element in it—the one picked out by the minimization of $\pi(\bar{y})$. Then from Eq. (5.20) we have that $E(\hat{e}, z) = E(\hat{e}, z - \bar{y})$, where z is any appropriate function. Choosing z to be the displacement when due to a point force at x_0, Eq. (5.25) becomes $|e(x_0)| \leqslant \|\hat{e}\|_m \|z - \bar{y}\|_m$, and if $\bar{y}$ can be selected to energetically approximate z so that $\|z - \bar{y}\|_m = O(h^q)$, $q > 0$, then a better bound than $\|\hat{e}\|$ is obtained for $|\hat{e}(x_0)|$. We notice in this connection that it matters greatly whether x_0 is inside an element or between two elements. In the first case, a discontinuity in y' inside the element may greatly reduce the best possible $\|z - \bar{y}\|_m$ while on the boundary between two elements it may not affect it at all. Take the fixed string with linear elements. If x_0 is at the nodes, then we have $\|z - \bar{y}\|_1 = 0$, resulting in $|\hat{e}(x)| = 0$, as we proved differently in Section 3 of this chapter.

6. Richardson's Extrapolation to the Limit

Knowledge of the theoretical rate of convergence can be used to improve the accuracy of the computed values through an extrapolation to $h = 0$ named after Richardson. Suppose the error is known to be $O(h^q)$ and that h is sufficiently small so that we may write

$$\hat{y} - y = ch^q + \varepsilon \tag{5.69}$$

with ε being much smaller than the dominant error term ch^q. Using two sets of computations for the two h values h_1 and h_2, we eliminate c in Eq. (5.69) and are left with

$$y = \frac{1}{h_2{}^q - h_1{}^q}(\hat{y}_1 h_2{}^q - \hat{y}_2 h_1{}^q) + \frac{1}{h_2{}^q - h_1{}^q}(\varepsilon_1 h_2{}^q - \varepsilon_2 h_1{}^q) \tag{5.70}$$

and if we neglect the supposedly small ε terms in Eq. (5.70), a better approximation for y is provided.

If q is uncertain, three numerical values may be used to determine it from

$$\log|y - \hat{y}| = \log c + q \log h \tag{5.71}$$

The success of Richardson's extrapolation hinges on the correctness of Eq. (5.71) or on the fact that the theoretical rate of convergence is reached. Computed $\hat{y}$ values that do not fall on the straight log–log line (5.71) may cause Richardson's extrapolation to *worsen* the accuracy of the extrapolated results. Figure 5.2 is an example for numerical results that fall neatly on a straight log–log line, but one has to beware of round-off errors that can confuse the picture. In practice, more than two $\hat{y}$ values are used for the Richardson extrapolation, and y in Eq. (5.71) is predicted after it has been ascertained that the straight log–log line (5.71) holds for them.

The computed central deflection of the circular clamped plate, point loaded at the center and discretized by a *uniform* mesh of cubic finite elements, is found in the numerical example of Section 2 to be $\hat{y}_1 = 0.8834$, $\hat{y}_2 = 0.9729$, and $\hat{y}_3 = 0.9874$, for $h_1 = 1$, $h_2 = \frac{1}{2}$, and $h_3 = \frac{1}{3}$, respectively. We profess to know that in this case the energy rate of convergence is only $O(h^2)$ and have with $q = 2$, $h_1 = 1$, $h_2 = \frac{1}{2}$, and neglecting ε in Eq. (5.69), that $c = -0.11933$, thus predicting that at $h = 0$, $y = 1.0027$ in place of the exact $y = 1$. As many as six finite elements would have been otherwise needed to obtain this accuracy. Feigning ignorance of the rate of convergence, we put the computed $\hat{y}$ values for $Ne = 1$, 2, and 3 into Eq. (5.71) and get that q is actually nearly 2, indicating that the origin is a singular point of order $r^2 \log r$, and suggesting an appropriately graded mesh of finite elements.

7. Numerical Integration

It may very well happen that $p(x)$, $q(x)$, and $f(x)$ are sufficiently complicated functions of x to warrant the numerical integration of $\pi(\tilde{y})$ in order to form the element matrices. Numerical integration is expensive and we are committed to keeping the number of sampling points in this integration to the minimum. The central question in that respect is what the minimum integration order is that will maintain the greatest accuracy of which the element is capable. To gain insight into this problem, consider $\pi(\tilde{y}) = \frac{1}{2}E(\tilde{y}, \tilde{y}) - (f, \tilde{y})$ with linear elements, necessitating the numerical integration of only $(f, \tilde{y})$. Gauss integration is most efficient on the computer. Suppose that one integration point taken at the center of each element, is employed to integrate $(f, \tilde{y})$. We shall specifically show that this integration scheme is sufficient to keep the original accuracy of the linear element. For that purpose, we regard the numerical integration as equivalent to the *replacement of the original $f(x)$ by a piecewise constant load $g(x)$*, as shown in Fig. 5.5. With this piecewise constant load the numerical integration of $\pi(\tilde{y})$ with one point is exact, but instead of solving the original problem, we now solve the neighboring *perturbed* problem. In this way, the error in the finite element

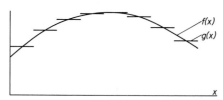

Fig. 5.5   Replacement of the smooth load function $f(x)$ by a piecewise constant function $g(x)$.

solution may be looked upon as consisting of the discretization or approximation errors plus the perturbation errors arising from the replacement of $f(x)$ by $g(x)$; we do not want the last to exceed the first. Since the discontinuities in $g(x)$ occur only on element boundaries, the analytic solution z to the perturbed problem is smooth within the elements, and the finite element solution *to the perturbed problem* is accurate $O(h)$ in the $\|\cdot\|_1$ norm. We hope that the perturbation error is also $O(h)$, in balance with the approximation error. The general analysis of this problem consists of studying the change in the solution of the boundary value problem resulting from changes in the coefficients of the differential equation. Here we restrict ourselves only to the case of changes in $f(x)$. Then let y be the solution to $-(py')' + qy = f$, and z the solution to $-(pz')' + qz = g$. The error $e = y - z$ is the solution of $-(pe')' + qe = f - g$, for which the equation of virtual displacements may be written as

$$E(e, e) = (e, f - g) \tag{5.72}$$

and consequently

$$\|e\|_1{}^2 \leqslant \|e\|_0 \|f - g\|_0 \tag{5.73}$$

We know from the analysis of Section 4 that $\|e\|_0 \leqslant c\|e\|_1$, with which Eq. (5.73) becomes $\|e\|_1 \leqslant c\|f - g\|_0$. With a piecewise constant $g(x)$, $\|f - g\|_0 = O(h)$ and hence $\|e\|_1 = O(h)$. The perturbation error caused by the numerical integration is in this case $O(h)$, in balance with the approximation error which is $O(h)$ too. This is the chief argument in the numerical integration error analysis. We shall not pursue it further except to recall the major theorem of this analysis. It states that *in order to maintain the full finite element accuracy in problems with variable coefficients, the numerical scheme should exactly integrate y'^2 in second-order problems and y''^2 in fourth-order problems.*

For instance, the elastic energies of a unit disk and a unit sphere are

$$E(y, y) = \int_0^1 \left(ry'^2 + \frac{1}{r} y^2 \right) dr, \qquad E(y, y) = \int_0^1 (r^2 y'^2 + 2y^2) \, dr \tag{5.74}$$

If discretized with polynomial finite elements of order p, both require an integration scheme that exactly integrates polynomials of degree $2p - 2$ or p Gauss points, the appearance of $1/r$ and r^2 in $E(y, y)$ notwithstanding.

EXERCISES

1. An elastic tube is carrying gas at pressure $p = p_0$. The operational formulation of this boundary value problem from the theory of elasticity

reads

$$\left(y' + \frac{1}{r}y\right)' = 0, \qquad \tfrac{1}{4} = r_0 < r < r_1 = 1$$

$$y'(r_0) = -p_0, \qquad y'(r_1) = 0$$

Associate with this problem the energy norm

$$\|y\|_1^2 = \int_{r_0}^{r_1} \left(ry'^2 + \frac{1}{r}y^2\right)dr$$

and establish its total potential energy. Discretize the problem with linear ($p = 1$) and quadratic ($p = 2$) finite elements. To form the linear element stiffness matrix, use Gauss integration with 1, 2, and then 3 points. Form the element stiffness matrix for the quadratic element by numerical integration with 2, 3, and then 4 Gauss points. In each case, solve the problem with 1, 2, 3, and 4 finite elements and draw $\hat{e}(r)$ and $\hat{e}'(r)$. Observe the error distribution in each case. Apply Richardson's extrapolation to the limit and compare your results with the correct

$$y = c_1 r + c_2\frac{1}{r}, \qquad c_1 = \frac{r_0^2}{r_1^2 - r_0^2}p_0, \qquad c_2 = -\frac{r_0^2 r_1^2}{r_1^2 - r_0^2}p_0$$

2. Consider a circular unit membrane whose deflection y is described by

$$-\frac{1}{r}\frac{d}{dr}\left(r\frac{dy}{dr}\right) = f(r), \qquad 0 < r < 1$$

$$y'(0) = 0, \qquad y(1) = 0$$

The total potential energy associated with the problem is

$$\pi(y) = \tfrac{1}{2}\int_0^1 (y'^2 - 2f(r)y)r\,dr$$

in which the trial function $\tilde{y}(r)$ has to satisfy the sole essential boundary condition $\tilde{y}(1) = 0$.

Now let the load $f(r)$ be a unit point force $P(0)$ at the origin such that

$$\int_0^1 P(0)yr\,dr = y(0)$$

The symbolic solution to this problem is

$$y = \log(1/r)$$

making no physical sense at $r = 0$, where $y(0) = \infty$, and being energetically implausible because

$$\|y\|_1^2 = \int_0^1 y'^2 r\,dr = \infty$$

with $y = \log(1/r)$.

For the particular case $f(r) = P(0)$, $\pi(y)$ becomes

$$\pi(y) = \tfrac{1}{2} \int_0^1 y'^2 r\, dr - y(0)$$

but $\tfrac{1}{2}\|y - \tilde{y}\|_1^2 = \pi(\tilde{y}) - \pi(y)$ makes no sense here, and $\pi(\tilde{y}) \geqslant \pi(y)$ from Eq. (3.22) reduces to $\pi(\tilde{y}) \geqslant -\infty$.

All this does not mean, however, that the mathematical model of a membrane loaded by a point load should be entirely discarded, nor that a useful finite element solution cannot be obtained from the last $\pi(y)$ for points removed from the singularity at $r = 0$. To see this, solve the problem of a point loaded membrane with linear finite elements and compare $\hat{y}(r)$ and $\hat{y}'(r)$ to $y(r) = \log(1/r)$ and $y'(r) = -1/r$ in the interval $0 < r \leqslant 1$ for decreasing mesh sizes.

Explore the effect of a nonuniform mesh in accelerating the reduction of $\pi(\tilde{y})$ with a mesh of the form $h_e = he^n$ for different values of $n \geqslant 0$, where the smallest element $h_1 = h$ is near the center of the membrane. Use also higher-order elements.

3. Because the response to a point load is infinite at the center of a circular membrane, a point central displacement given to the unloaded membrane is not transmitted. In other words the two-point boundary value problem

$$(ry')' = 0 \quad 0 < r < 1 \qquad y(0) = 1, \quad y(1) = 0$$

has no solution. In fact, here $y = c_1' \log r + c_2$ and c_1 and c_2 cannot be adjusted to make y satisfy the boundary conditions. The total potential energy is here

$$\pi(y) = \frac{1}{2} \int_0^1 ry'^2\, dr \qquad y(0) = 1, \quad y(1) = 0$$

and $\pi(y) \geqslant 0$. Show that with $y = 1 - r^\varepsilon$, $\varepsilon > 0$ we can bring $\pi(y)$ as close to zero as we wish by taking ε sufficiently small. Minimize $\pi(y)$ with linear finite elements and observe the behavior of the computed solution.

SUGGESTED FURTHER READING

Fox, L., *The Numerical Solution of Two Point Boundary Problems in Ordinary Differential Equations.* Oxford Univ. Press (Clarendon), London and New York, 1957.

Kantorovich, L. V., and Krylov, V. I., *Approximate Methods of Higher Analysis.* Wiley (Interscience), New York, 1958.

Strang, G., and Fix, G. J., *An Analysis of the Finite Element Method.* Prentice–Hall, Englewood Cliffs, New Jersey, 1973.

Oden, J. T., and Reddy, J. N., *An Introduction to the Mathematical Theory of Finite Elements.* Wiley, New York, 1976.

6 Eigenproblems

1. Stability of Columns

An intuitively graspable eigenproblem borrowed from the field of *elastic stability*, appropriate to describe the gist of this kind of problem, is that of the equilibrium of a thin elastic beam or column, axially compressed, as shown in Fig. 6.1. The column shown in the figure is held in a slightly bent position by the axial end forces P. These forces build a moment $M = Py$ on the beam's middle line, and according to the elementary theory of beams, which assumes $y \ll l$, the beam's reaction is

$$M = \pm p(x)y'' \tag{6.1}$$

where $p(x)$ includes the elastic and geometrical properties of the beam's cross section. The sign ambiguity in Eq. (6.1) is resolved through the observation that when y is negative, y'' is positive and vice versa. The equation of equilibrium of the axially compressed columns is thus found to be

$$p(x)y'' + Py = 0, \qquad 0 < x < l$$
$$y(0) = y(l) = 0 \tag{6.2}$$

A notable aspect of the boundary value problem (6.2), the one that actually causes it to be an *eigenproblem*, is that it is *homogeneous*, being satisfied by the *trivial* solution $y = 0$. Physically, the trivial solution $y = 0$ is utterly

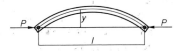

Fig. 6.1 Axially compressed thin elastic beam hinged at both ends.

plausible because if we assume the situation to be ideal (a perfectly straight beam acted on only by axial forces), then the originally flat beam should, if undisturbed, remain in its straight equilibrium position for any P. But are there also, perhaps for some particular values of P, nontrivial solutions to the homogeneous boundary value problem (6.2) such that the column may assume a bent shape after being slightly perturbed? To simplify the search for these solutions, we assume that the beam is of homogeneous composition and geometrically uniform so that $p(x) = 1$, and write Eq. (6.2) in the simpler form

$$y'' + \omega^2 y = 0, \qquad 0 < x < l, \qquad P = \omega^2$$
$$y(0) = y(l) = 0$$
(6.3)

The problem of locating the values of ω^2 for which $y \neq 0$ is a solution to Eq. (6.3) constitutes an eigenproblem. Here, the general solution to Eq. (6.3) is rather simple:

$$y = c_1 \sin \omega x + c_2 \cos \omega x \tag{6.4}$$

which still needs to be made to satisfy the boundary conditions $y(0) = y(l) = 0$. The condition $y(0) = 0$ is taken care of with $c_2 = 0$, but to avoid the trivial solution $y = 0$, we refrain from setting $c_1 = 0$. Instead, we seek out those values of ω for which $y(1) = 0$. This occurs when

$$\sin \omega l = 0 \tag{6.5}$$

or

$$\omega l = n\pi, \qquad n = 1, 2, \ldots, \infty \tag{6.6}$$

such that

$$P_n = (\pi n/l)^2, \qquad n = 1, 2, \ldots, \infty \tag{6.7}$$

When the axial force P equals one of the *specific* values in Eq. (6.7), the beam can assume a bent or *buckled* equilibrium configuration. The smallest such force, $P_{\mathrm{crit}} = (\pi/l)^2$, is of special importance in structural design. Corresponding to $P_n = (n\pi/l)^2$ is the buckling *mode* $y_n = \sin(n\pi x/l)$, shown in Fig. 6.2, which has arbitrary amplitude. The particular values of ω^2, usually denoted by λ, for which a nontrivial solution exists for the boundary value problem (6.3) are the *eigenvalues* of the problem while the corresponding nontrivial solutions $y = \sin \omega x$ are the *eigenfunctions* or buckling modes of the column. Corresponding to the *fundamental* eigenvalue $\lambda_1 = P_{\mathrm{crit}}$ is the eigenfunction $y = \sin \pi x/l$. All eigenfunctions are determined only up to the multiplying factor c_1, but in forming Eq. (6.3) we assumed that y is much less than 1. In fact, the buckling modes are only the *incipient* state of the column at the moment its equilibrium configuration may change from the straight

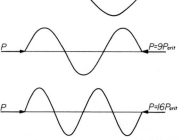

Fig. 6.2   Buckling modes of the beam in Fig. 6.1.

to the bent position. What happens later at *postbuckling* must be described by a nonlinear equation. We shall not omit this equilibrium phase, but defer its analysis to Section 11.

So long as $P < P_{crit}$, a small perturbation imparted to the column, such as with a light transverse tug, encounters restoring forces that tend to return the column back to its straight equilibrium position, Above P_{crit}, a disturbance in the perfect symmetry of the beam or the load may cause it to snap into a buckled position, and the column may even collapse.

Shown in Fig. 6.1 is a column that is hinged, or simply supported at both its ends, and the absence of a bending moment, which is proportional to y'' at these points, is assumed by Eq. (6.1), for which $y'' = 0$ when $y = 0$. Clamped or built-in ends call for all the boundary conditions $y(0) = y'(0) = y(1) = y'(1) = 0$ to be explicitly imposed on the eigenfunction y. Four coefficients are needed in the general solution of the boundary value problem for that to occur. We can make these coefficients appear by raising the order of the eigenproblem from two to four, so that now

$$(p(x)y'')'' + Py'' = 0, \qquad 0 < x < l$$
$$y(0) = y'(0) = y(l) = y'(l) = 0 \tag{6.8}$$

In the event of a constant $p(x) = p$, and with $P/p = \omega^2$, Eq. (6.8) becomes

$$y'''' + \omega^2 y'' = 0, \qquad 0 < x < l \tag{6.9}$$

possessing the general solution

$$y = c_1 + c_2 x + c_3 \cos \omega x + c_4 \sin \omega x$$
$$y' = c_2 - c_3 \omega \sin \omega x + c_4 \omega \cos \omega x \tag{6.10}$$

that is made to satisfy the boundary conditions $y(0) = y'(0) = y(l) = y'(l) = 0$ through

$$c_1 + c_3 = 0$$

$$c_2 + \omega c_4 = 0$$

$$c_1 + c_2 l + c_3 \cos \omega l + c_4 \sin \omega l = 0 \tag{6.11}$$

$$c_2 - c_3 \omega \sin \omega l + c_4 \omega \cos \omega l = 0$$

Choosing to concisely write $c = \cos \omega l$ and $s = \sin \omega l$, together with some eliminations, reduces Eq. (6.11) to

$$c_1(1 - c) + c_2[l - (1/\omega)s] = 0$$

$$c_1 \omega s + c_2(1 - c) = 0 \tag{6.12}$$

To secure a nontrivial solution y, the case $c_1 = 0$ and $c_2 = 0$ is to be avoided in Eq. (6.12), the determinant of the two equations needs to vanish, and we have from this that

$$2(1 - c) = \omega l s \tag{6.13}$$

which factors into the two equations

$$\sin \tfrac{1}{2} \omega l = 0, \qquad \tfrac{1}{2} \omega l = \tan \tfrac{1}{2} \omega l \tag{6.14}$$

with first roots that yield

$$P_{\text{crit}} = p(2\pi/l)^2 \quad \text{and} \quad P_{\text{crit}} = p(2.86\pi/l)^2 \tag{6.15}$$

respectively. The first of Eqs. (6.14) is readily solved for one set of eigenvalues: $\omega^2 = n^2(2\pi/l)^2$, for which $\sin \omega l = 0$ and $\cos \omega l = 1$. Consequently, $c_2 = c_4 = 0$ and this set of eigenfunctions (buckling modes) of Eq. (6.9) with the boundary conditions $y(0) = y'(0) = y(l) = y'(l)$ is

$$y = 1 - \cos \omega x \tag{6.16}$$

As before, also now, there are infinitely many eigenfunctions and as many corresponding eigenvalues.

A more general fourth-order eigenvalue problem that describes the buckling of an elastically supported column, axially compressed, is

$$(p(x)y'')'' + q(x)y + Py'' = 0, \qquad 0 < x < l \tag{6.17}$$

with appropriate boundary conditions.

In addition to physical significance, the eigenfunctions possess some important intrinsic properties, as will become evident soon, to warrant their computation.

2. Vibration of Elastic Systems

Another technical field in which eigenproblems are prominent is vibration of elastic bodies. A simple example of this is the taut string fixed at its ends, held at tension p, and of a (possibly variable) density $\rho(x)$ per unit of its length. To create the string's equation of motion, we apply Newton's second law of motion to a small segment of the string as shown in Fig. 6.3, and have that in the y direction

$$\rho\ddot{y}\,\delta x = -p\sin\theta + p\sin(\theta + \delta\theta) \tag{6.18}$$

in which the double overdots indicate second-order differentiation with respect to the time t. To *linearize* Eq. (6.18), we assume small deflections such that θ can replace $\sin\theta$, and upon proceeding to the limit $\delta x \to 0$, we get the equation

$$\rho(x)\ddot{y} = py'' \tag{6.19}$$

which is a *partial* differential equation, y being a function of both x and t. When $\rho(x)$ is the constant ρ, Eq. (6.19) is written in full as

$$\frac{\partial^2 y}{\partial t^2} = \frac{p}{\rho}\frac{\partial^2 y}{\partial x^2} \tag{6.20}$$

Fig. 6.3   Deflected string under the tension p.

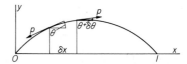

The solution of this *partial* differential equation can be reduced to that of an *ordinary* eigenproblem by *separation of variables* in which the space–time problem is separated into independent problems in space and time. In this, the solution to Eq. (6.20) is attempted in the form $y(x,t) = y(x)e^{i\omega t}$, such that

$$y'' = y''(x)e^{i\omega t}, \qquad \ddot{y} = -\omega^2 y(x)e^{i\omega t} \tag{6.21}$$

since $i^2 = -1$. From Eqs. (6.19) and (6.21) we learn that $y(x)$ has to satisfy the homogeneous equation

$$-\omega^2\rho(x)y = py'' \tag{6.22}$$

which is only one dimensional. For a constant $\rho(x) = \rho$ and fixed ends, Eq. (6.22) for $y(x)$ takes the form

$$y'' + \lambda y = 0, \qquad \lambda = \omega^2\rho/p, \qquad 0 < x < l$$
$$y(0) = y(l) = 0 \tag{6.23}$$

which we recognize to be the first eigenproblem solved in Section 1. Eigen-problem (6.23) is known to have the eigenvalues $\lambda_1 = (\pi/l)^2, \lambda_2 = (2\pi/l)^2, \dots,$ $\lambda_n = (n\pi/l)^2$ with the corresponding eigenfunctions $y_n = \sin\sqrt{\lambda_n}x$, or when $\rho/p = 1$, $y_n = \sin\omega_n x$. The solution to Eq. (6.20) is then the sum

$$y(x, t) = \sum_{j=1}^{\infty} y_j(x)(\alpha_j \cos\sqrt{\lambda_j}t + \beta_j \sin\sqrt{\lambda_j}t) \tag{6.24}$$

in which the coefficients α_j and β_j are determined by the initial string con-figuration $y(x, 0)$ and the initial velocity distribution $\dot{y}(x, 0)$. Equation (6.24) describes the movement of the string in space and time in terms of the eigenfunctions of Eq. (6.23), which are the *natural vibration modes* $y_1(x)$, $y_2(x), \dots, y_n(x)$ of the string, and also in terms of the *natural frequencies* $\omega_1, \omega_2, \dots, \omega_n = (n\pi/l)\sqrt{p/\rho}$ of the same string.

A major incentive for computing the natural frequencies and modes of the string, or for that matter of any other elastic body, rests with Eq. (6.24). Another important phenomenon in the mechanics of vibrations, requiring natural frequencies for its prediction and explanation, is *resonance*. Suppose the string is being acted upon by a *periodic* force of the form $f(x)e^{i\Omega t}$; then

$$-py'' + \rho(x)\ddot{y} = f(x)e^{i\Omega t} \tag{6.25}$$

Separation of variables works also for this nonhomogeneous case if we write $y(x, t) = y(x)e^{i\Omega t}$, such that

$$y'' = y''(x)e^{i\Omega t}, \qquad \ddot{y} = -\Omega^2 y(x)e^{i\Omega t} \tag{6.26}$$

Substitution of y'' and $\ddot{y}$ from Eq. (6.26) into the partial differential equation (6.25) accomplishes its reduction to the ordinary boundary value problem

$$y'' + \Omega^2 y = -f, \qquad 0 < x < l$$
$$y(0) = y(l) = 0 \tag{6.27}$$

in which, for the sake of simplicity, we assumed $\rho = p = 1$. We may further assume, without loosing any of the example's demonstrative quality, that f in Eq. (6.27) is constant. Then the general solution to this equation is

$$y = c_1 \cos\Omega x + c_2 \sin\Omega x - (f/\Omega^2) \tag{6.28}$$

The constants c_1 and c_2 in this equation are decided by the boundary con-ditions $y(0) = y(l) = 0$ and when fulfilled,

$$y(x) = \frac{f}{\Omega^2 \sin\Omega l}[\sin\Omega x - \sin\Omega l + \sin\Omega(l - x)] \tag{6.29}$$

reaching a maximum at $x = l/2$. At this point

$$y\left(\frac{l}{2}\right) = \frac{f}{\Omega^2} \frac{1 - \cos\frac{1}{2}\Omega l}{\cos\frac{1}{2}\Omega l} \tag{6.30}$$

and when $\Omega = \pi/l$—the fundamental natural frequency of the string—it resonates; $y(l/2) \to \infty$.

Resonance is an eminently undesirable incident fraught with possibly disasterous consequences. It may occur, for instance, when a poorly balanced piece of machinery rhythmically shakes the elastic structure on which it is mounted, causing annoying, if not outrightly damaging, high amplitude vibrations. To anticipate and take precautionary measures against resonance, the natural frequencies of the elastic foundation have to be computed inside the range within which the frequency of the exciting periodic force is expected.

A more general second-order eigenproblem, that for the elastically supported string, is

$$-(p(x)y')' + q(x)y = \lambda\rho(x)y \tag{6.31}$$

often named after *Sturm–Liouville*.

A convenient way to write down the proper equation of motion of an elastic body is offered by *D'Alembert's principle*, which calls for the replacement of the static load $f(x)$ in the boundary value problem by the inertia load $-\rho\ddot{y}$. Applying this principle to the elastically supported beam, we readily deduce from Eq. (2.29) the beam's equation of motion

$$(p(x)y'')'' + q(x)y = -\rho(x)\ddot{y} \tag{6.32}$$

where $\rho(x)$ is density per unit length. Equation (6.32) is accompanied by appropriate boundary and initial conditions. Separation of variables converts the partial differential equation (6.32) to the ordinary eigenvalue problem

$$(p(x)y'')'' + q(x)y = \lambda\rho(x)y \tag{6.33}$$

with the same boundary conditions.

3. Finite Difference Approximation

Technically, the approximation of an eigenvalue problem, like that in Eq. (6.23), by finite differences is the same as that of a boundary value problem.

Following the procedure of Chapter 1, we divide the string into $N + 1$ sections which for convenience we take to be of equal size $h = 1/(N + 1)$ and label the nodes thus created sequentially from 0 to $N + 1$. For the approximate replacement of y'', we choose the central difference scheme (1.14), using which, we write at all interior points the following finite difference equations

$$h^{-2}(y_0 - 2y_1 + y_2) + \lambda y_1 = 0$$
$$h^{-2}(y_1 - 2y_2 + y_3) + \lambda y_2 = 0$$
$$\vdots \qquad\qquad (6.34)$$
$$h^{-2}(y_{N-1} - 2y_N + y_{N+1}) + \lambda y_N = 0$$

After the boundary conditions $y_0 = y_{N+1} = 0$ have been introduced into Eq. (6.34), and with the usual notation $y^T = (y_1, y_2, \ldots, y_N)$, the homogeneous system of algebraic equations (6.34) is compactly written as

$$(K - \lambda I)y = 0 \qquad\qquad (6.35)$$

in which K is the matrix

$$K = \frac{1}{h^2} \begin{bmatrix} 2 & -1 & & & & \\ -1 & 2 & -1 & & & \\ & \cdot & \cdot & \cdot & & \\ & & \cdot & \cdot & \cdot & \\ & & & \cdot & \cdot & \\ & & & -1 & 2 & -1 \\ & & & & -1 & 2 \end{bmatrix} \qquad\qquad (6.36)$$

that previously appeared in the string boundary value problem.

The *algebraic* eigenproblem (6.34) is the discrete counterpart to the continuous one in Eq. (6.23). Its solution supplies us with N approximate eigenvalues $\lambda_1 \leqslant \lambda_2 \leqslant \cdots \leqslant \lambda_N$ and N corresponding *eigenvectors* $y_1, y_2, \ldots, y_N$. Computational procedures to solve the algebraic eigenproblem have been devised and analyzed at length and in depth within the context of computational linear algebra. Efficient and reliable routines for this purpose are included in most computer libraries for convenient use.

To estimate the discretization errors that beset the eigenvector y and the eigenvalue λ, we make use of the regular pattern in the difference equations (6.34) to elicit a closed form formula for y and λ that we shall compare with the exact values.

A typical equation from among those in system (6.34) is written as

$$-y_{j-1} + (2 - \lambda h^2)y_j - y_{j+1} = 0 \qquad\qquad (6.37)$$

which is observed to admit the solution

$$y_j = cz^j \tag{6.38}$$

where z is a constant determined by Eq. (6.37). Introduction of $y_j = cz^j$ into Eq. (6.37) occasions the *auxilliary* or *characteristic* equation

$$z^2 - (2 - \lambda h^2)z + 1 = 0 \tag{6.39}$$

associated with the finite *difference* equation in the same way it is associated with the *differential* equation. The two roots of Eq. (6.39) are

$$z_{1,2} = \tfrac{1}{2}[(2 - \lambda') \pm \sqrt{\lambda'(\lambda' - 4)}], \qquad \lambda' = \lambda h^2 \tag{6.40}$$

Gerschgorin's theorem assures us that for K in Eq. (6.36), $0 \leqslant \lambda' \leqslant 4$. For the limiting case $\lambda' = 0$ or $\lambda' = 4$, the two roots z_1 and z_2 of Eq. (6.39) are equal, according to Eq. (6.40), to $z = (2 - \lambda')/2$ and the solution to the difference equation (6.37) takes the form

$$y_j = (c_1 + c_2 j)z^j \tag{6.41}$$

where c_1 and c_2 are determined from the boundary conditions $y_0 = y_{N+1} = 0$. But these boundary conditions are satisfied only when $c_1 = c_2 = 0$ and a nontrivial condition y does not exist for $\lambda' = 0$ or $\lambda' = 4$; they are not eigenvalues of K in Eq. (6.36). We conclude that $0 < \lambda' < 4$, and hence from Eq. (6.40), that $z_{1,2}$ are complex conjugates which we prefer to write as $z_1 = ce^{i\theta}$ and $z_2 = ce^{-i\theta}$. Since the last term in Eq. (6.39) is 1, we have from Descartes' rule that $z_1 z_2 = 1$, which means that $c = 1$, $z_1 = e^{i\theta}$, and $z_2 = e^{-i\theta}$. The general solution to Eq. (6.37) is therefore $y_j = c_1 e^{ij\theta} + c_2 e^{-ij\theta}$, or since $e^{i\theta} = \cos\theta + i\sin\theta$,

$$y_j = c_1 \cos j\theta + c_2 \sin j\theta \tag{6.42}$$

Because $y_0 = 0$, $c_1 = 0$ and one boundary condition is met. Satisfaction of the other boundary condition $y_{N+1} = 0$ requires that

$$\sin(N+1)\theta = 0, \qquad \theta = n\pi h, \qquad n = 1, 2, \ldots, N \tag{6.43}$$

if we want to evade the trivial case $c_2 = 0$, and hence

$$(y_j)_n = c_2 \sin \pi jnh \tag{6.44}$$

which is the complete expression for the jth entry in the nth eigenvector. It is determined up to the multiplicative constant c_2 that we conveniently take to be 1.

We recall that the exact eigenfunctions of the fixed string are $y_n(x) = \sin n\pi x$, and hence according to Eq. (6.44), *at the nodes* $(y_j)_n = y_n(x_j)$. At these points the discrete values are exact.

With the eigenvectors secured, we turn to the computation of the approximate eigenvalues, which we obtain from the characteristic equation (6.39) by putting $z = e^{i\theta}$ back into it to get

$$\lambda h^2 = 2 - (e^{i\theta} + e^{-i\theta}) = 2(1 - \cos\theta) \tag{6.45}$$

Or, since $1 - \cos\theta = 2\sin^2\frac{1}{2}\theta$, the approximate nth eigenvalue λ_n is given by the formula

$$\lambda_n = (4/h^2)\sin^2\tfrac{1}{2}n\pi h, \qquad h = 1/(N+1) \tag{6.46}$$

At the lower end of the *discrete spectrum* where $\frac{1}{2}n\pi h \ll 1$, $\sin\theta \cong \theta$ and approximately

$$\lambda_n = n^2\pi^2(1 - \tfrac{1}{12}n^2\pi^2 h^2) \tag{6.47}$$

But $n^2\pi^2$ are the exact eigenvalues to which the approximate λ_n in Eq. (6.47) match, and the approximation error in the eigenvalues is nearly $-n^2\pi^2 h^2/12$, or $O(n^2 h^2)$ when $n \ll N$. As we climb in the spectrum, the relative accuracy of λ_n declines but it still converges $O(h^2)$.

No more is said on finite differences since finite elements will soon appear to be a highly systematic method for the discretization of the eigenvalue problem given in its variational formulation. General error estimates are also possible for the finite element approximation of the eigenvalue problem in the same way they were estimated in Chapter 4 for the boundary value problem.

4.   Rayleigh's Quotient

Rayleigh's quotient is central to the finite element discretization of the eigenvalue problem. We present it and its basic properties in this section, immediately after we have discovered some of the more salient properties of the eigenvalues and eigenfunctions of the problems we are engaged in solving.

Let $\bar{y}$ be a trial function acceptable in the principle of minimum potential energy. Multiplication of $-(p(x)y')' + q(x)y = \lambda\rho(x)y$ by this function and integration by parts lead us to the equation of virtual displacements for the eigenproblem

$$E(y, \bar{y}) = \lambda K(y, \bar{y}) \tag{6.48}$$

in which $K(y, \bar{y})$ is the more general inner product

$$K(y, \bar{y}) = \int_0^1 \rho(x)y\bar{y}\,dx \tag{6.49}$$

where $\rho(x) \geqslant 0$. The letter K is used in Eq. (6.49) for the fact that $K(y, y)$ is proportional to the momentary *kinetic* energy of the string. Henceforth we shall use $K(\,,)$ instead of the particular $(\,,)$ inner product.

Of great interest is the property of two eigenfunctions y_n and y_m, corresponding to two *distinct* eigenvalues λ_n and λ_m, that they are at the same time orthogonal and energy orthogonal. To prove this we write Eq. (6.48) twice, once for $y = y_m$ and then for $y = y_n$, with y taken to be first y_n and then y_m, and have from all this that

$$E(y_m, y_n) = \lambda_m K(y_m, y_n)$$
$$E(y_n, y_m) = \lambda_n K(y_n, y_m)$$

(6.50)

Inasmuch as the eigenproblems we deal with are symmetric, $E(y, z) = E(z, y)$, $K(y, z) = K(z, y)$, and the subtraction of Eqs. (6.50) from each other yields

$$(\lambda_m - \lambda_n)K(y_m, y_n) = 0$$

(6.51)

When $\lambda_m \neq \lambda_n$, it follows from Eqs. (6.50) and (6.51) that

$$K(y_m, y_n) = 0, \qquad E(y_m, y_n) = 0$$

(6.52)

The fact that the eigenfunctions are orthogonal with respect to both $K(\,,)$ and $E(\,,)$, plus the proven fact that when $\rho(x) > 0$ they are *complete* with respect to $\|\cdot\|_m$—any energetically admissible function can be expressed, energetically as closely as desired, as a linear combination of these eigenfunctions—makes them ideal basis functions for the Ritz method. Such basis functions are readily seen to create *diagonal* global stiffness and mass matrices. Unfortunately, in problems with variable coefficients, the computation of the eigenfunctions is by far more expensive and involved than that of the static solution.

The energies we are dealing with are bounded from below with a positive constant such that

$$E(y, y) \geqslant \omega^2 K(y, y)$$

(6.53)

holds for any function y that satisfies the essential boundary conditions, and for which $E(y, y) < \infty$. Let ω_1 be the highest possible ω in Eq. (6.53) and y_1 the function that goes along with it, such that

$$E(y_1, y_1)/K(y_1, y_1) = \omega_1{}^2$$

(6.54)

Let y be an energetically admissible function with which $y_1 + \varepsilon y$ is also energetically admissible (i.e., satisfying the essential boundary conditions and with a finite energy). The ratio

$$R[y_1 + \varepsilon y] = E(y_1 + \varepsilon y, y_1 + \varepsilon y)/K(y_1 + \varepsilon y, y_1 + \varepsilon y)$$

(6.55)

is minimal at $\varepsilon = 0$, and $dR/d\varepsilon = 0$ produces the equation

$$E(y_1, y)K(y_1, y_1) - E(y_1, y_1)K(y_1, y) = 0 \qquad (6.56)$$

which with $E(y_1, y_1)/K(y_1, y_1) = \omega_1^2$ becomes

$$E(y_1, y) - \omega_1^2 K(y_1, y) = 0 \qquad (6.57)$$

which is recognized to be the equation of virtual displacements for the eigenvalue problem. It holds true for *any* energetically admissible y, and hence by the calculus of variations argument of Chapter 3, Section 5, y_1 is the eigenfunction corresponding to $\lambda_1 = \omega_1^2$.

Consider, for instance, the fixed string for which

$$E(y_1, y) = \int_0^1 y_1' y' \, dx, \qquad K(y_1, y) = \int_0^1 y_1 y \, dx \qquad (6.58)$$

Substituting these into Eq. (6.57), integrating by parts, and with the boundary conditions $y(0) = y_1(0) = y(1) = y_1(1) = 0$, we get that

$$\int_0^1 y(y_1'' + \omega_1^2 y_1) \, dx = 0 \qquad (6.59)$$

for any admissible y. The fundamental theorem of the calculus of variations asserts that a necessary condition for this equality to hold for *any* such y is that $y_1'' + \omega_1^2 y_1 = 0$.

Rayleigh's quotient is defined as

$$R[y] = E(y, y)/K(y, y) \qquad (6.60)$$

where y is an energetically admissible function. From Eqs. (6.53) and (6.54) we reveal the first of its major properties, namely,

$$R[y] \geqslant \lambda_1 \qquad (6.61)$$

with equality occurring only when $y = y_1$.

More generally

$$R[y] \geqslant \lambda_n, \qquad K(y, y_1) = K(y, y_2) = \cdots = K(y, y_{n-1}) = 0 \qquad (6.62)$$

which is proved by selecting a trial function in the form

$$y = z - \sum_{j=1}^{n-1} K(z, y_j)y_j, \qquad K(y_j, y_j) = 1 \qquad (6.63)$$

in which z is arbitrary, y_j is the jth eigenfunction; and this trial function satisfies the required orthogonalities in Eq. (6.62).

Inequality (6.62) establishes a variational formulation for the eigenvalue problem that we write as

$$\lambda_n = \min_y R[y], \qquad y \perp y_1 \perp y_2 \perp \cdots \perp y_{n-1} \qquad (6.64)$$

This is the celebrated Rayleigh theorem or principle. In Eq. (6.64), the minimization is carried out over the class of all functions y that are energetically admissible in the principle of minimum potential energy.

Lord Rayleigh himself (in 1870 and later in 1899) put his quotient, given in Eq. (6.60), to use in a manner that served as the forerunner to the later Ritz (Rayleigh–Ritz to the English) method for the estimation of the fundamental (lowest) natural frequencies of various elastic bodies. As a quick example of Rayleigh's technique, we compute the fundamental frequency of the fixed string of length l, constant density ρ, and tension p. On physical grounds, we choose the trial function $y = x(x - l)$ that correctly satisfies the essential boundary conditions $y(0) = y(1) = 0$. For this string

$$R[y] = \frac{p}{\rho} \int_0^l y'^2 \, dx \bigg/ \int_0^l y^2 \, dx \qquad (6.65)$$

and

$$R[y] = 10 \frac{p}{\rho l^2} \geq \lambda_1 = \pi^2 \frac{p}{\rho l^2} \qquad (6.66)$$

which is a close approximation indeed.

It is not by chance that such a good approximation for λ_1 is computed with Rayleigh's quotients from a rather crude approximation to y_1. Rayleigh's quotient is inherently accurate for the eigenvalues. To elicit this other important property of $R(y)$, we assume the approximation to the nth eigenfunction y_n of the form $y = y_n + \varepsilon z$, with both y_n and z *normalized* with respect to the kinetic energy $K(\ , \ )$, such that $K(y_n, y_n) = K(z, z) = 1$. Then, when $|\varepsilon| \ll 1$,

$$R[y] = \lambda_n + \varepsilon^2 [E(z, z) - \lambda_n] \qquad (6.67)$$

and an error $O(\varepsilon)$ in y_n induces, with Rayleigh's quotient, an error only $O(\varepsilon^2)$ in λ_n.

5. Finite Element Approximation

Discretization of the eigenvalue problem with finite elements is achieved through the introduction of the piecewise polynomial trial function $\tilde{y}$ into Rayleigh's quotient and its minimization with respect to the nodal values. As in boundary value problems, also here, for a typical element

$$K_e(\tilde{y}, \tilde{y}) = y_e^T m_e y_e, \qquad E_e(\tilde{y}, \tilde{y}) = y_e^T k_e y_e \qquad (6.68)$$

in which y_e is the element nodal values vector, k_e the element stiffness matrix, and m_e the element mass matrix. Actually, the term "mass matrix" has its origin in Eq. (6.68), in which m_e is the generalized mass in the kinetic energy expression $y_e^T m_e y_e$ of the eth element. The element matrices are created from the shape functions $\phi_1, \phi_2, \ldots, \quad \phi_n$ by

$$(k_e)_{ij} = E_e(\phi_i, \phi_j), \qquad (m_e)_{ij} = K_e(\phi_i, \phi_j) \qquad (6.69)$$

Their assembly has been thoroughly discussed in Chapter 4 and need not be repeated. In their assembled form, the elastic and kinetic energies are

$$K(\tilde{y}, \tilde{y}) = y^T M y, \qquad E(\tilde{y}, \tilde{y}) = y^T K y \qquad (6.70)$$

with which Rayleigh's quotient becomes

$$R[\tilde{y}] = y^T K y / y^T M y \qquad (6.71)$$

where K and M are the global stiffness and mass matrices [the *matrix* K is not to be confused with $K(\ ,\ )$] and y is the global vector containing the nodal displacements. The best approximation of λ_1 is obtained from Eq. (6.71) by the condition that $R(\tilde{y})$ be minimal or $\partial R[\tilde{y}]/\partial y = 0$, which produces the equation

$$\frac{2}{y^T M y}\left(K y - \frac{y^T K y}{y^T M y} M y \right) = 0 \qquad (6.72)$$

With linearly independent shape functions and $\rho > 0$, m_e is positive definite. Consequently, so is M, $y^T M y > 0$, if $y \neq 0$, and Eq. (6.72) takes the form

$$K y = \lambda M y, \qquad \lambda = y^T K y / y^T M y \qquad (6.73)$$

Essential boundary conditions, if there are any, need yet be enforced upon $\tilde{y}$. Their inclusion in Eq. (6.73) is entirely analogous to that for the discretized boundary value problem. For instance, if one of the entries of y, say y_j, is supressed: $y_j = 0$, then $\partial R[\tilde{y}]/\partial y_j = 0$, and the jth row in Eq. (6.73) must be deleted. Also, since $y_j = 0$, the jth column in both K and M can be filled with zeros too. At this point, either the algebraic eigenproblem is compressed after removing from K and M the jth row and column, or the jth row and column in both K and M is set equal to zero and M_{jj} is set equal to 1. This last established practice is attractive to the programmer but produces a spurious eigenvalue $\lambda = 0$ that we have to know to ignore.

For the fixed string of constant density ρ and tension p, discretized with first-order finite elements, we have from Eq. (4.23) that

$$k_e = \frac{p}{h}\begin{bmatrix} 1 & -1 \\ -1 & 1 \end{bmatrix}, \qquad m_e = \frac{\rho h}{6}\begin{bmatrix} 2 & 1 \\ 1 & 2 \end{bmatrix} \qquad (6.74)$$

assembled into

$$\left\{ \frac{p}{h^2} \begin{bmatrix} 2 & -1 & & & \\ -1 & 2 & -1 & & \\ & & \ddots & \ddots & \ddots & \\ & & & -1 & 2 & -1 \\ & & & & -1 & 2 \end{bmatrix} - \lambda \frac{\rho}{6} \begin{bmatrix} 4 & 1 & & & \\ 1 & 4 & 1 & & \\ & & \ddots & \ddots & \ddots & \\ & & & 1 & 4 & 1 \\ & & & & 1 & 4 \end{bmatrix} \right\} y = 0 \quad (6.75)$$

in which the essential boundary conditions $y(0) = y(1) = 0$ were introduced through the removal of the first and last rows and columns in both K and M.

Comparison of the matrices in Eq. (6.75), produced with finite elements, with those in Eqs. (6.35) and (6.36), produced by finite differences, reveals that in both cases it is the same K, but M is different. While the finite difference method sets up a diagonal M matrix, that produced by the finite element method has the same nonzero entries distribution as K. The first apparent reward we get for accepting the added complexity stemming from the finite element discretization is that λ_1, computed with it, is always *above* the exact fundamental eigenvalue. Using the minmax principle of the next section, we shall be able to show that this is true for *all* the eigenvalues computed with finite elements from Rayleigh's quotient; they are all above the exact ones. This does not mean that finite elements deliver more accurate eigenvalues less expensively. In fact, the accuracy of λ_1 computed from the finite difference system (6.35), and that computed from the finite element system (6.75) is the same $O(h^2)$. In practice, the added information provided by the bounds on the finite element eigenvalues is of limited benefit, particularly in light of the fact that *lower* bounds on the eigenvalues are nearly impossible to compute.

Solution of the generalized algebraic eigenproblem $Ky = \lambda My$ is, as we know, more time and computer-space consuming than that of the particular $Ky = \lambda y$. For this reason, we shall investigate in Section 9 ways to modify the finite element technique to produce a diagonal, or *lumped*, mass matrix, thereby relinquishing the bounds provision but, hopefully, without loosing the systematics and efficiency of the finite element method.

6.   The Minmax Principle

This principle is derived from Rayleigh's but is somewhat subtler and we approach it with circumspection. We start arguing the minmax principle

by requiring that the trial function y in Rayleigh's quotient be spanned by
any n eigenfunctions, such that

$$y = c_1 y_i + c_2 y_j + \cdots + c_n y_m \tag{6.76}$$

Let $\lambda_i \leqslant \lambda_j \leqslant \cdots \leqslant \lambda_m$ be the eigenvalues corresponding to $y_i, y_j, \ldots, y_m$.
Rayleigh's theorem affirms that with the y in Eq. (6.76)

$$\lambda_m = \max_{c_1, \ldots, c_n} R[y] \tag{6.77}$$

If this maximization is repeated for *all possible combinations* of n eigenfunctions, as in Eq. (6.76), then the *lowest* value we shall ever encounter for
λ_m is λ_n, which occurs for the choice $y = c_1 y_1 + c_2 y_2 + \cdots + c_n y_n$. Hence, a
characterization of the nth eigenvalue λ_n is possible in the form

$$\lambda_n = \min\{\max R[y]\} \tag{6.78}$$

in which the maximization is over $c_1, c_2, \ldots, c_n$ for a *particular* choice of n
eigenfunctions in Eq. (6.76), and the minimization is over all the possible
linear combinations of n eigenfunctions. Equation (6.78) expresses the
minmax principle, except that in its general formulation, $y_i, y_j, \ldots, y_m$ in
Eq. (6.76) may be *any n* linearly independent and energetically admissible
functions, not just the eigenfunctions. To generalize the minmax principle,
it is crucial that we show that for *any choice* of the *linearly independent*
functions $y_i, y_j, \ldots, y_m$ in Eq. (6.76), there exists a set of coefficients
$c_1, c_2, \ldots, c_n$ for which $R[y] \geqslant \lambda_n$. The proof of this hinges on the fact that
with at least one of these coefficients not equal to zero y can be made orthogonal to the first $n - 1$ eigenfunctions $y_1, y_2, \ldots, y_{n-1}$. Indeed, to satisfy

$$K(y, y_1) = K(y, y_2) = \cdots = K(y, y_{n-1}) = 0 \tag{6.79}$$

at most $n - 1$ coefficients are used up in Eq. (6.76) from among the n available, and so at least one of the cs can be assured to be nonzero. If so, then by
Rayleigh's principle, a function y that is orthogonal to $y_1, y_2, \ldots, y_{n-1}$
satisfies the inequality $R[y] \geqslant \lambda_n$ and we obviously have that

$$\max_{c_1, \ldots, c_n} R[y] \geqslant \lambda_n, \qquad y = c_1 y_i + c_2 y_j + \cdots + c_n y_m \tag{6.80}$$

$y_i, y_j, \ldots, y_m$ being any n linearly independent energetically admissible
functions. Equality occurs in Eq. (6.80) when y is the *eigenfunction* corresponding to λ_n, and hence we have Eq. (6.78).

From here on, we shall regularly have to distinguish between the exact
eigenvalue λ and the corresponding one computed with finite elements
through the minimization of Rayleigh's quotient. Consistent with the
notation of Chapter 4, we shall denote the latter by $\hat{\lambda}$.

A similar minmax principle can be developed for the finite element function space $\tilde{y}$. There

$$\hat{\lambda}_n = \min\{\max R[\tilde{y}]\} \tag{6.81}$$

where

$$\tilde{y} = c_1\tilde{y}_1 + c_2\tilde{y}_2 + \cdots + c_n\tilde{y}_n \tag{6.82}$$

in which $\tilde{y}_1, \tilde{y}_2, \ldots, \tilde{y}_n$ are any n linearly independent finite element functions picked out from the space of $\tilde{y}$, and where the maximization is carried out over the weights $c_1, c_2, \ldots, c_n$. Minimization in Eq. (6.81) is carried out over all the possible choices of $\tilde{y}_1, \tilde{y}_2, \ldots, \tilde{y}_n$. From Eq. (6.80), we have that $\max R[\tilde{y}] \geqslant \lambda_n$, but since the search for this maximum is restricted to $\tilde{y}$, the equality $R[\tilde{y}] = \lambda_n$ may not occur, and we reason that

$$\hat{\lambda}_n \geqslant \lambda_n, \qquad n = 1, 2, \ldots, N \tag{6.83}$$

All finite element eigenvalues $\hat{\lambda}_n$ are above the corresponding exact ones. It should not escape our attention, though, that inequality (6.83) assumes strict adherence to the variational formulation of the eigenproblem. Approximate numerical integration of $E(y, y)$ or $K(y, y)$ will negate it and the bound will be forfeited.

7. Discretization Accuracy of Eigenvalues

The goal of this section is to estimate $|\lambda_n - \hat{\lambda}_n|$ in terms of the discretization and intrinsic parameters of the problem the way we did in Chapter 5 for the boundary value problem. In general, it is easier to accomplish this estimate for λ_1 than for λ_n. We open our discussion on the finite element error analysis of eigenproblems with an estimate for $|\lambda_1 - \hat{\lambda}_1|$. The general error analysis will come later.

Let y_1 be the eigenfunction corresponding to the lowest exact eigenvalue λ_1 and $\tilde{y}$ a proper finite element trial function. Also let $e(x) = \tilde{y}(x) - y_1(x)$. Since $R[\tilde{y}] \geqslant \lambda_1$, we have that

$$\hat{\lambda}_1 \leqslant \frac{E(y_1, y_1) + 2E(y_1, e) + E(e, e)}{1 + 2K(y_1, e) + K(e, e)} \tag{6.84}$$

in which it is assumed that y_1 is normalized; $K(y_1, y_1) = 1$. When use is made of $E(y_1, y_1) = \lambda_1$, Eq. (6.84) becomes

$$\hat{\lambda}_1 - \lambda_1 \leqslant \frac{E(e, e) - \lambda_1 K(e, e)}{1 + 2K(y_1, e) + K(e, e)} \tag{6.85}$$

The Cauchy–Schwarz inequality assures us that

$$|K(y_1,e)| \leqslant ||e||_0 = K^{1/2}(e,e) \tag{6.86}$$

and we have from this that

$$1 + 2K(y_1,e) + K(e,e) \geqslant 1 - 2||e||_0 + ||e||_0^2 = (1 - ||e||_0)^2 \tag{6.87}$$

Since $-\lambda_1 K(e,e) < 0$, $E(e,e) = ||e||_m^2$, and $\hat{\lambda}_1 - \lambda_1 \geqslant 0$, we may write Eq. (6.85) in the form

$$|\lambda_1 - \hat{\lambda}_1| \leqslant ||e||_m^2/(1 - ||e||_0)^2 \tag{6.88}$$

and have, when $||e||_0 \ll 1$, that the error in the fundamental eigenvalue λ_1, computed with finite elements, is essentially the energy error in the trial function for y_1. Remembering the results of Chapter 5, we obtain that in the eigenproblem of the 2mth order, discretized with polynomial finite elements of order p,

$$|\lambda_1 - \hat{\lambda}_1| \leqslant ch^{2(p+1-m)} \max|y_1^{(p+1)}|^2 \tag{6.89}$$

For the linear ($p = 1$) string ($m = 1$) element, Eq. (6.89) predicts that $|\lambda_1 - \hat{\lambda}_1| = O(h^2)$, the same accuracy that we previously had with the central three point finite difference scheme for y''.

To estimate the accuracy of the higher eigenvalues, we enlist the minmax principle from which we have that

$$\hat{\lambda}_n \leqslant \max_{c_1,\ldots,c_n} R[\tilde{y}], \qquad \tilde{y} = c_1\tilde{y}_1 + c_2\tilde{y}_2 + \cdots + c_n\tilde{y}_n \tag{6.90}$$

where $\tilde{y}_1, \tilde{y}_2, \ldots, \tilde{y}_n$ are any n linearly independent finite element functions from the same space $\tilde{y}$. We write the errors in the exact eigenfunctions $y_1, y_2, \ldots, y_n$ as

$$e_1 = \tilde{y}_1 - y_1, \qquad e_2 = \tilde{y}_2 - y_2, \qquad \ldots, \qquad e_n = \tilde{y}_n - y_n \tag{6.91}$$

and seek to bound $|\lambda_n - \hat{\lambda}_n|$ in terms of $e_1, e_2, \ldots, e_n$. For this, we use Eq. (6.91) to write y in Eq. (6.90) as

$$\tilde{y} = (c_1 y_1 + c_2 y_2 + \cdots + c_n y_n) + (c_1 e_1 + c_2 e_2 + \cdots + c_n e_n) \tag{6.92}$$

or simply

$$\tilde{y} = y + e \tag{6.93}$$

Furthermore, we decompose e into a component e^l spanned by the eigenfunctions $y_1, y_2, \ldots, y_n$ and another one $e^\perp$ which is orthogonal to $y_1, y_2, \ldots, y_n$; and may now write Eq. (6.93) as

$$\tilde{y} = y + e^l + e^\perp \tag{6.94}$$

Fig. 6.4　Schematic of relationships among the function y spanned by the eigenfunctions $y_1, y_2, \ldots, y_n$, the finite element trial function $\tilde{y}$, the error $e = \tilde{y} - y$, and the orthogonal components e^l and $e^\perp$ of e.

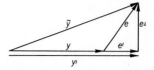

Since both y and e^l are spanned by $y_1, y_2, \ldots, y_n$, we combine them into y^l and write

$$\tilde{y} = y^l + e^\perp \tag{6.95}$$

Of course, $E(y^l, e^\perp) = K(y^l, e^\perp) = 0$, and the vector diagram in Fig. 6.4 graphically summarizes the other relation ships among y, y^l, $\tilde{y}$, e, $e^\perp$, and e^l.

When $\tilde{y}$ in Eq. (6.95) is put into Rayleigh's quotient in Eq. (6.90) it becomes

$$\hat{\lambda}_n \leqslant \frac{E(y^l, y^l) + E(e^\perp, e^\perp)}{K(y^l, y^l) + K(e^\perp, e^\perp)} \tag{6.96}$$

where, for typographical brevity, the "max" has been dropped from before $R[\tilde{y}]$, assuming that the $c_1, c_2, \ldots, c_n$ coefficients in $\tilde{y}$ are the ones that maintain the inequality. We are already assured by the minmax principle that $\hat{\lambda}_n \geqslant \lambda_n$, and further have from Eq. (6.96) that

$$|\lambda_n - \hat{\lambda}_n| \leqslant \frac{1}{K(\tilde{y}, \tilde{y})} [E(y^l, y^l) - \lambda_n K(y^l, y^l) + E(e^\perp, e^\perp) - \lambda_n K(e^\perp, e^\perp)] \tag{6.97}$$

By the fact that y^l is spanned by the first n eigenfunctions, we have from Rayleigh's theorem that

$$E(y^l, y^l) - \lambda_n K(y^l, y^l) \leqslant 0 \tag{6.98}$$

and these terms, together with the other negative term $-\lambda_n K(e^\perp, e^\perp)$, may be dropped from the righthand side of inequality (6.97), leaving us with

$$|\lambda_n - \hat{\lambda}_n| \leqslant E(e^\perp, e^\perp)/K(\tilde{y}, \tilde{y}) \tag{6.99}$$

But

$$E(e, e) = E(e^l, e^l) + E(e^\perp, e^\perp) \tag{6.100}$$

and consequently

$$E(e, e) \geqslant E(e^\perp, e^\perp) \tag{6.101}$$

Also, by the triangle inequality

$$K(\tilde{y}, \tilde{y}) = \|\tilde{y}\|_0^2 = \|y + e\|_0^2 \geqslant (\|y\|_0 - \|e\|_0)^2 = K(y, y)\left(1 - \frac{\|e\|_0}{\|y\|_0}\right)^2 \tag{6.102}$$

with which Eq. (6.98) becomes

$$|\lambda_n - \hat{\lambda}_n| \leqslant E(e, e)/K(y, y)(1 - \|e\|_0/\|y\|_0)^2 \tag{6.103}$$

We recall that $e = c_1 e_1 + c_2 e_2 + \cdots + c_n e_n$ and proceed to get

$$e^2 \leqslant (c_1{}^2 + c_2{}^2 + \cdots + c_n{}^2)(e_1{}^2 + e_2{}^2 + \cdots + e_n{}^2) \tag{6.104}$$

and then

$$\|e\|_0^2 \leqslant \sum_{j=1}^{n} c_j{}^2 \sum_{j=1}^{n} \|e_j\|_0^2 \tag{6.105}$$

If the eigenfunctions system $y_1, y_2, \ldots, y_N$ is assumed orthonormal, such that $K(y_j, y_j) = 1$ and $E(y_i, y_j) = 0$, $i \neq j$, then since $y = c_1 y_1 + c_2 y_2 + \cdots + c_n y_n$, $\|y\|_0^2 = c_1{}^2 + c_2{}^2 + \cdots + c_n{}^2$ Eq. (6.105) becomes

$$\frac{\|e\|_0}{\|y\|_0} \leqslant \left(\sum_{j=1}^{n} \|e_j\|_0^2 \right)^{1/2} \tag{6.106}$$

A reasonable assumption on $\tilde{y}_1, \tilde{y}_2, \ldots, \tilde{y}_n$ is that they are good approximations, say the interpolates to the eigenfunctions $y_1, y_2, \ldots, y_n$, to the extent that $\|e\|_0/\|y\|_0 \ll 1$, with which Eq. (6.103) simplifies into

$$|\lambda_n - \hat{\lambda}_n| \leqslant E(e, e)/K(y, y) \tag{6.107}$$

Application of the triangle inequality to $E^{1/2}(e, e)$, where $e = c_1 e_1 + c_2 e_2 + \cdots + c_n e_n$, produces the inequality

$$E(e, e) \leqslant \sum_{j=1}^{n} c_j{}^2 \sum_{j=1}^{n} E(e_j, e_j) \tag{6.108}$$

or

$$E(e, e) \leqslant K(y, y) \sum_{j=1}^{n} E(e_j, e_j) \tag{6.109}$$

and the bound in Eq. (6.107) becomes

$$|\lambda_n - \hat{\lambda}_n| \leqslant \sum_{j=1}^{n} \|e_j\|_m^2 \tag{6.110}$$

which fulfills our objective.

The functions $\tilde{y}_1, \tilde{y}_2, \ldots, \tilde{y}_n$ in Eq. (6.82) are arbitrary. Taken to be the interpolates to $y_1, y_2, \ldots, y_n$, respectively, the error analysis of Chapter 5, Section 2, establishes that

$$\|e_j\|_m^2 \leqslant ch^{2(p+1-m)} |y_j^{(p+1)}|^2 \tag{6.111}$$

Our eigenproblems are essentially of the form $y'' = \lambda y$ and $y'''' = \lambda y$, or in general, $y^{(2m)} = \lambda y$ and $y^{(2mq)} = \lambda y^{(q)} = \lambda^q y$, such that $|y^{(p+1)}| = c\lambda^{(p+1)/(2m)}$, therefore

$$\|e_j\|_m^2 \leq ch^{2(p+1-m)}\lambda_j^{(p+1)/m} \tag{6.112}$$

Finally

$$|\lambda_n - \hat{\lambda}_n| \leq ch^{2(p+1-m)} \sum_{j=1}^n \lambda_j^{(p+1)/m} \tag{6.113}$$

The sum from 1 to n in Eq. (6.113) seems to be an overestimation, for if we choose $\tilde{y}_1, \tilde{y}_2, \ldots, \tilde{y}_n$ in Eq. (6.90) to be the interpolates to $y_1, y_2, \ldots, y_n$, then, at least with a fine mesh maximization will choose $c_n \cong 1$, with the rest of the coefficients small and $|\lambda_n - \hat{\lambda}_n| \cong E(e_n, e_n)$, in lieu of the sum. Without this sum, the error estimate for the eigenvalues takes the form

$$|1 - (\hat{\lambda}_n/\lambda_n)| \leq ch^{2(p+1-m)}\lambda_n^{(p+1-m)/m}, \qquad n \ll N \tag{6.114}$$

predicting that for the linear ($p = 1$) string ($m = 1$) element $|1 - (\hat{\lambda}_n/\lambda_n)| \leq O(\lambda_n h^2)$ as in Eq. (6.47).

To substantiate experimentally the theoretical error estimate in Eq. (6.114), we undertake to compute the natural frequencies of a cantilever beam, entailing the discretization and solution of $y'''' = \lambda y$, $0 < x < 1$ with $y(0) = y'(0) = y''(1) = y'''(1) = 0$. The first five eigenfunctions, or natural modes, of this beam are shown in Fig. 6.5. Cubic C^1 beam elements are employed to discretize the beam with the element mass and stiffness matrices taken from Eqs. (4.47) and (4.53), respectively. Here $m = 2$, $p = 3$, and Eq. (6.114) foresees that $|1 - (\hat{\lambda}_n/\lambda_n)| = O(\lambda_n h^4)$. Figure 6.6 shows the convergence of the first

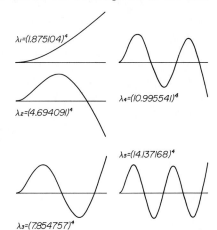

Fig. 6.5 First five natural modes of vibration and the corresponding eigenvalues of a cantilever beam.

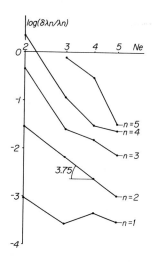

Fig. 6.6   Relative error $\delta\lambda_n/\lambda_n = 1 - \hat{\lambda}_n/\lambda_n$ in the first five eigenvalues of the cantilever beam discretized with cubic finite elements versus the number of finite elements Ne.

five eigenvalues versus the number of finite elements Ne employed in the discretization. Extraction of the eigenvalues from the algebraic system was done with a Jacobi method on a computer with about 6.5 significant digits. This helps to explain the seemingly erratic convergence of λ_1. The discretization errors in λ_1, which *decrease* with Ne, are already at the *round-off* errors level, which *increase* with Ne, and hence the interference. Convergence of λ_2 is nearly as predicted by Eq. (6.114); the numerical convergence rate being 3.75 instead of the theoretical 4. As for the higher modes, the condition $n \ll N$ in Eq. (6.114) does not hold for them any more, and the curves lack a discernible constant slope.

Figure 6.7 shows the accuracy of the first five eigenvalues in the case of a *fixed mesh* composed of five cubic elements. Equation (6.114) predicts the error in λ_n to be proportional to itself, and indeed the curve λ_n versus n drawn in Fig. 6.7, and adjusted to pass through the experimental value for $n = 2$, also passes accurately through the points for $n = 3$ and $n = 4$. The numerical result for $n = 1$ is off the theoretical prediction, possibly because of round-off error interference, and that for $n = 5$, possibly because $n = 5$ is too high.

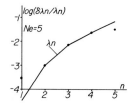

Fig. 6.7   Relative error $\delta\lambda_n/\lambda_n = 1 - \hat{\lambda}_n/\lambda_n$, as a function of n, in a cantilever beam discretized with five cubic finite elements. Dots mark numerical results.

8. Discretization Accuracy of Eigenfunctions

With an error estimate secured for the eigenvalues, we are in a position to proceed to estimate the accuracy of the eigenfunctions computed with finite elements. Let $\hat{y}_1$ be the computed finite element eigenfunction approximating the fundamental mode y_1. If we imagine y_1 and $\hat{y}_1$ to be normalized vectors, then they differ only in direction and this *directional error* θ_1 is schematically shown in Fig. 6.8. Assume, then, that both y_1 and $\hat{y}_1$ are normalized and write

$$\hat{y}_1 = y_1 + e_1 = y_1 + e_1{}^{|} + e_1{}^{\perp} = y_1{}^{|} + e_1{}^{\perp} \tag{6.115}$$

It is seen in Fig. 6.8 that

$$K(e^{\perp}, e^{\perp}) = \sin^2 \theta_1 \tag{6.116}$$

and

$$\|y_1 - \hat{y}_1\|_0^2 = 4 \sin^2 \tfrac{1}{2}\theta_1 \tag{6.117}$$

To express the error in the eigenfunction in terms of the error in the eigenvalues, we write

$$\hat{\lambda}_1 - \lambda_1 = E(\hat{y}_1, \hat{y}_1) - \lambda_1 \tag{6.118}$$

which, since $E(y_1{}^{|}, y_1{}^{|}) = \lambda_1 K(y_1{}^{|}, y_1{}^{|})$, yields

$$\hat{\lambda}_1 - \lambda_1 = E(e_1{}^{\perp}, e_1{}^{\perp}) - \lambda_1 K(e_1{}^{\perp}, e_1{}^{\perp}) \tag{6.119}$$

Furthermore, because $e^{\perp}$ is orthogonal to y_1, we have from Rayleigh's theorem that

$$E(e_1{}^{\perp}, e_1{}^{\perp}) \geqslant \lambda_2 K(e_1{}^{\perp}, e_1{}^{\perp}) \tag{6.120}$$

and consequently, Eq. (6.119) becomes

$$\hat{\lambda}_1 - \lambda_1 \geqslant (\lambda_2 - \lambda_1) K(e^{\perp}, e^{\perp}) \tag{6.121}$$

Recalling Eq. (6.116), we finally get the desired bound

$$\sin^2 \theta_1 \leqslant |\lambda_1 - \hat{\lambda}_1| / (\lambda_2 - \lambda_1) \tag{6.122}$$

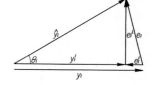

Fig. 6.8 Schematic diagram of the first exact and computed eigenfunctions y_1 and $\hat{y}_1$, both normalized, and the error $e_1 = \hat{y}_1 - y_1$ decomposed into $e^{\perp}$ orthogonal to y_1 and $e^{|}$ parallel to it.

It can be shown that for the higher eigenfunctions

$$(\lambda_{n+1} - \lambda_n)\sin^2\theta_n \leqslant \hat{\lambda}_n - \lambda_n + \sum_{n=1}^{n-1} (\lambda_{n+1} - \lambda_n)\sin^2\theta_n \qquad (6.123)$$

where θ_n is the angle between y_n and $\hat{y}_n$.

Extrapolation to the limit discussed in Chapter 5, Section 6, for boundary value problems can equally well be applied to eigenproblems in order to improve the accuracy of both eigenvalues and eigenfunctions. The accompanying table lists the first five eigenvalues of the cantilever beam,

Mode	$Ne = 2$	$Ne = 3$	Limit	Exact
1	12.374180	12.365582	12.363466	12.362354
2	493.79248	488.70825	487.45675	485.51831
3	5648.5703	3901.9941	3472.0684	3806.5449
4	47584.086	19788.398	12946.394	14617.270
5		70088.312		39943.824

computed with cubic ($p = 3$) elements. Computation was carried out with two and three finite elements and extrapolation to the limit was made under the assumption that $\hat{\lambda} \cong \lambda + cNe^{-4}$. We observe in this table that for λ_1 and λ_2 the extrapolation improves the accuracy. The results for λ_3 and λ_4 are less desirable and are also below the exact values. But with no more than three elements in the discretization, we do not yet expect λ_3 and λ_4 to follow closely enough the asymptotic prediction $\hat{\lambda} - \lambda = ch^4$.

9.   Change of Basis: Condensation

One of the main conclusions we draw from the error analysis of Section 8 is that in order to improve the accuracy of any *fixed* eigenvalue λ_n, the mesh of finite elements needs to be refined and the algebraic problem increased in size. In practice, only the lower portion of the eigenvalue spectrum is of interest, as say, in stability analysis, for which only the fundamental mode and eigenvalue is desired. Nevertheless, to achieve an acceptable accuracy in these eigenvalues, a considerable number of elements is usually needed, resulting in sizable matrices. Fulfilling the need, algorithms have been devised for the computation of only the first, or first few, eigenvalues when the rest of them is of no interest; but owing to the size of the matrices, these algorithms can be fairly expensive.

The reason for the large number of elements needed to approximate y_1 is that the finite element basis functions with their compact support are, individually, ill suited to describe y_1. It is a certain disadvantage of finite elements compared with the classical Ritz method, in which the larger freedom in the selection of the basis functions permits us to choose only a few of them to adaquately represent y_1. This suggests the profitable combination of the two methods: The global mass and stiffness matrices are first assembled, using the systematics of finite elements, but then their size is reduced or *condensed* through a proper selection of a more restricted basis, richer in the lower eigenfunctions, such that fewer of these basis functions are needed for the desired accuracy of the lower spectrum.

To describe condensation, let the algebraic eigenproblem be $Ky = \lambda My$ with K and M of dimension N. Let $z_1, z_2, \ldots, z_n$, $n < N$, be the new vector basis such that

$$y = c_1 z_1 + c_2 z_2 + \cdots + c_n z_n = Zc \qquad (6.124)$$

Then by substitution and premultiplication by Z^T

$$Z^T K Z c = \lambda Z^T M Z c \qquad (6.125)$$

which is of dimension n only.

As an example of condensation, take the fixed string discretized with finite differences, resulting in the algebraic eigenproblem in Eqs. (6.35) and (6.36). Let h in this example be $\frac{1}{8}$. We wish to condense the 7×7 algebraic eigenproblem into a 3×3 one. To this end, we choose the three new basis functions, or vectors, shown in Fig, 6.9, for which the matrix Z in Eq. (6.125) is

$$Z = \begin{bmatrix} 1 & 3 & 1 \\ 2 & 6 & 2 \\ 3 & 5 & 3 \\ 4 & 4 & 4 \\ 3 & 3 & 5 \\ 2 & 2 & 6 \\ 1 & 1 & 3 \end{bmatrix} \qquad (6.126)$$

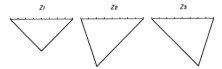

$z_1 \qquad\qquad z_2 \qquad\qquad z_3$

Fig. 6.9 Three global basis functions for the condensation of the string eigenproblem.

and consequently, the system in Eq. (6.125) becomes

$$\left\{\begin{bmatrix} 1 & 1 & 1 \\ 1 & 3 & 1 \\ 1 & 1 & 3 \end{bmatrix} - \lambda \frac{1}{2}h^2 \begin{bmatrix} 11 & 15 & 15 \\ 15 & 25 & 19 \\ 15 & 19 & 25 \end{bmatrix}\right\} c = 0 \qquad (6.127)$$

yielding

$$\lambda_1 = 10.1133, \qquad \lambda_2 = 42.6667, \qquad \lambda_3 = 95.2960 \qquad (6.128)$$

instead of

$$\lambda_1 = 9.7434, \qquad \lambda_2 = 37.4903, \qquad \lambda_3 = 79.0164 \qquad (6.129)$$

from the original 7×7 system.

With only three elements

$$\left\{\begin{bmatrix} 2 & -1 & \\ -1 & 2 & -1 \\ & -1 & 2 \end{bmatrix} - \frac{1}{16}\lambda \begin{bmatrix} 1 & & \\ & 1 & \\ & & 1 \end{bmatrix}\right\} y = 0 \qquad (6.130)$$

yielding

$$\lambda_1 = 9.3726, \qquad \lambda_2 = 32.00, \qquad \lambda_3 = 54.63 \qquad (6.131)$$

All these approximate eigenvalues correspond to the exact

$$\lambda_1 = 9.8696, \qquad \lambda_2 = 39.4784, \qquad \lambda_3 = 88.8264 \qquad (6.132)$$

The three eigenfunctions derived from the condensed system in Eq. (6.126) are pictured in Fig. 6.10.

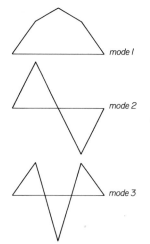

mode 1

mode 2   *Fig. 6.10* Three modes computed from the condensed string eigenproblem (6.127).

mode 3

Notice that all three $\lambda_1, \lambda_2, \lambda_3$ in Eq. (6.128), computed after condensation, are *above* the $\lambda_1, \lambda_2, \lambda_3$ in Eq. (6.129) in accordance with the prediction of the minmax principle.

10. Numerical Integration: Lumping

For the eigenproblem, as for the boundary value problem, the study of the effect of perturbation in the stiffness and mass coefficients in the differential equation upon the eigenvalues leads to the conclusion that the exact integration, of y'^2 in second-order problems and y''^2 in fourth-order problems is sufficient to preserve the full rate of convergence inherent in the polynomial shape functions.

Take, for example, the unit elastic disk for which

$$E(y, y) = \int_0^1 \left(ry'^2 + \frac{1}{r} y^2 \right) dr, \qquad K(y, y) = \int_0^1 y^2 r \, dr \qquad (6.133)$$

Notwithstanding the r and $1/r$ in the integrands, if the shape functions include a polynomial of degree p, then a numerical integration of degree $2p - 2$ is sufficient, for both $E(\tilde{y}, \tilde{y})$ and $K(\tilde{y}, \tilde{y})$, to sustain the full rate of convergence. If a Gauss integration method is used, then with polynomial shape functions of degree p, p Gauss points suffice to assure the $O(h^{2p})$ accuracy in the computed eigenvalues. Figure 6.11 shows the convergence of the first eigenvalue λ_1 in the elastic disk ($m = 1$) discretized with quadratic ($p = 2$) finite elements. Two Gauss points are sufficient to maintain the $O(h^4)$ accuracy in λ_1, as confirmed in this figure.

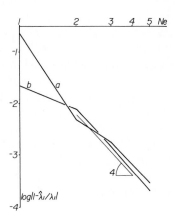

Fig. 6.11 Convergence of the computed fundamental eigenvalue $\hat{\lambda}_1$ of an elastic disk fixed at its rim and discretized with quadratic finite elements versus the number of elements Ne along the radius. Curves (a) and (b) refer to the numerical integration of k_e and m_e with two and three Gauss points, respectively.

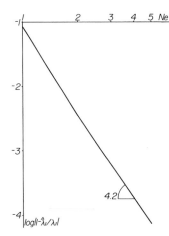

Fig. 6.12 Convergence of $\hat{\lambda}_1$ in a simply supported elastic plate discretized with cubic elements of which k_e and m_e were formed by using a two-point Gauss integration.

For the axisymmetrical circular plate,

$$E(y, y) = \int_0^1 \left(ry''^2 + \frac{1}{r} y'^2 \right) dr, \qquad K(y, y) = \int_0^1 ry^2 \, dr \qquad (6.134)$$

and again r and $1/r$ in the integrands of E and K notwithstanding, no more that $p - 1$ Gauss points are needed to integrate both $E(\tilde{y}, \tilde{y})$ and $K(\tilde{y}, \tilde{y})$, when the shape functions for $\tilde{y}$ are polynomials of degree p, to preserve the full rate of convergence. This is substantiated in Fig. 6.12 which shows the convergence $O(Ne^4)$ of the first eigenvalue λ_1 of the circular plate, discretized with Ne cubic elements and with the element matrices integrated with two Gauss points.

One of the more interesting applications of numerical integration is to best diagonalize or *lump* the element, and consequently, the global finite element mass matrix. The motivation for exactly integrating $E(\tilde{y}, \tilde{y})$ and $K(\tilde{y}, \tilde{y})$ is twofold: to obtain the upper bound on the exact eigenvalues guaranteed by the minmax principle, and to be assured of the theoretical rate of convergence predicted in Eq. (6.114). But there is a steep computational price to pay for these theoretical certitudes. The *consistent* mass matrix created in the orthodox application of the finite element method is non-diagonal and the algebraic eigenproblem stemming from it is of the general form $(K - \lambda M)y = 0$ rather than the (much cheaper to store and solve) special one $(K - \lambda I)y = 0$ generated with finite differences. We are tempted, then, to do away with the exact variational formulation, renounce its upper bound provision, and use instead numerical integration of a degree just sufficient to uphold the full rate of convergence, but, at the same time, sufficient to produce a *diagonal* element mass matrix.

How this is done is best demonstrated on concrete examples. Consider first the string with linear finite elements. To preserve the $O(h^2)$ accuracy of the eigenvalues, only constants need be exactly integrated in both $E(y, y)$ and $K(y, y)$. We fix the integration points at the nodes of this element and write

$$\int_0^h \tilde{y}^2 \, dx = \tfrac{1}{2}h(y_1{}^2 + y_2{}^2) \tag{6.135}$$

which is certainly correct for a constant $\tilde{y}$. The element mass matrix corresponding to Eq. (6.135) is

$$m_e = \frac{1}{2}h\begin{bmatrix} 1 & \\ & 1 \end{bmatrix} \tag{6.136}$$

which, when assembled into the global M, produces essentially the same mass matrix as does finite differences.

A more suggestive example of lumping by numerical integration is found in the cubic ($p = 3$) string element of Fig. 6.13. This element has four nodal points: two external and two internal, and an accuracy $O(h^6)$ is implicit in the cubic shape functions. Since here $\tilde{y} = x^3$ and $\tilde{y}'^2 = x^4$, the numerical integration scheme must exactly integrate polynomials of degree four. To have a diagonal m_e, we need to place the integration points at the nodes. The two end nodal points are fixed but we may vary the location of the internal ones, as well as the integration weights belonging to them, to attain the desired integration accuracy. This integration order is achieved with the weights $m_1 = \tfrac{1}{6}$, $m_2 = \tfrac{5}{6}$, and the internal integration points placed at $c = \pm\sqrt{5}/5$ as in the *Lobatto* integration method. For an element of size h,

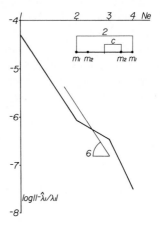

Fig. 6.13 Cubic string element with its mass lumped at the nodes. Convergence of $\hat{\lambda}_1$ computed with this element is at the optimal rate.

we get from all this that the lumped element mass matrix is

$$m_e = \frac{h}{12} \begin{bmatrix} 1 & & & \\ & 5 & & \\ & & 5 & \\ & & & 1 \end{bmatrix} \tag{6.137}$$

The above m_e has been used to compute the fundamental eigenvalue λ_1 of a fixed string and the results drawn in Fig. 6.13, where it is plain that the full accuracy $O(Ne^{-6})$ is kept with lumping.

Lumping of mass in the beam element, whose nodal values include slopes, is different. Cubic shape functions call for the exact integration quadratics. To exactly integrate $K(y, y)$ when $y = 1$ and $y = x$, we adjust the integration weights, or nodal masses, to get

$$m_e = \frac{h}{12} \begin{bmatrix} 6 & & & \\ & -h^2 & & \\ & & 6 & \\ & & & -h^2 \end{bmatrix} \tag{6.138}$$

which, unfortunately, includes *negative* masses, an occurrence that is often computationally detrimental.

11.   Nonlinear Eigenproblems

The critical buckling force P_{crit} of the axially compressed thin elastic column discussed in Section 1 was seen to be only the incipient value of the axial force at the very moment the column loses stability. In its bent position, the column assumes a too greatly displaced form for the assumption of infinitesimal displacements [inherent in the *linear* eigenproblem (6.2)] to be still valid, and it fails to describe properly the *postbuckling* behavior of the column. Beyond buckling, we have to accept the nonlinear equation of equilibrium

$$\phi'' + \lambda \sin \phi = 0, \qquad 0 < x < 1$$
$$\phi'(0) = \phi'(1) = 0 \tag{6.139}$$

in which ϕ is the angle between the middle line of the flexed column and the thrust axis and where x runs along the bent middle line. From this equation, we compute the amplitude of deflection in its dependence on the axial thrust λ.

Discretization of Eq. (6.139), using finite differences, does not differ from the same discretization in the linear case, and for five sampled ϕs at equal intervals h, we get the algebraic s stem

$$\frac{1}{h^2}\begin{bmatrix} 1 & -1 & & & \\ -1 & 2 & -1 & & \\ & -1 & 2 & -1 & \\ & & -1 & 2 & -1 \\ & & & -1 & 2 \end{bmatrix}\begin{bmatrix} \phi_1 \\ \phi_2 \\ \phi_3 \\ \phi_4 \\ \phi_5 \end{bmatrix} = \lambda \begin{bmatrix} \frac{1}{2}\sin\phi_1 \\ \sin\phi_2 \\ \sin\phi_3 \\ \sin\phi_4 \\ \sin\phi_5 \end{bmatrix}, \qquad h = \frac{1}{10} \quad (6.140)$$

in which a symmetric buckled shape has been assumed for the column. We propose to write this system as

$$\phi_2 = \phi_1 - \tfrac{1}{2}\lambda h^2 \sin\phi_1$$
$$\phi_3 = -\phi_1 + 2\phi_2 - \lambda h^2 \sin\phi_2$$
$$\phi_4 = -\phi_2 + 2\phi_3 - \lambda h^2 \sin\phi_3 \qquad (6.141)$$
$$\phi_5 = -\phi_3 + 2\phi_4 - \lambda h^2 \sin\phi_4$$
$$0 = -\phi_4 + 2\phi_5 - \lambda h^2 \sin\phi_5$$

If λ is fixed in Eq. (6.141), an arbitrary $\phi_1 > 0$ is chosen, and ϕ_2, ϕ_3, ϕ_4, and ϕ_5 are computed from these equations, then the last will not be fulfilled, except by chance, leaving the residual

$$r = -\phi_4 + 2\phi_5 - \lambda h^2 \sin\phi_5 \qquad (6.142)$$

This is repeated for different ϕ_1 values and the corresponding r is traced against ϕ_1 as in Fig. 6.14, where it is done for $\lambda = 16$. We expect the r versus ϕ_1 curve to pass through the origin for any $\lambda > 0$ since $\phi_1 = 0$ causes all other ϕs to be zero, which is the trivial solution to Eq. (6.139). But, in our case, where $\lambda = 16$, $r = 0$ also at $\phi_1 = 1.875$, corresponding to a nontrivial solution of Eq. (6.139). Choosing different λs, we get in this way different nonzero ϕ_1 values, drawn one versus the other in Fig. 6.15. Below the *bifurcation point* $\lambda = P_{\text{crit}}$, the column can hold itself in the sole equilibrium

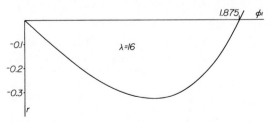

Fig. 6.14 Residual r versus the maximum slope ϕ_1 from Eqs. (6.141) and (6.142).

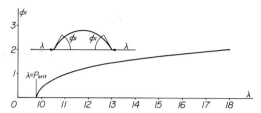

Fig. 6.15   Pre- and postbuckling dependence of ϕ_1 upon λ.

configuration $\phi(x) = 0$. At postbuckling, when $\lambda > P_{crit}$ for *any* value of axial force, the column possesses a bent equilibrium configuration and ϕ_1 increases as λ does, as shown in Fig. 6.15. Notice that if $\phi(x)$ is a nontrivial solution to the column equation (6.139), so is $\phi(x) + 2\pi j, j = 1, 2, 3, \ldots$.

Instead of the computational procedure we used above to solve the nonlinear eigenproblem, we may apply to this problem the method of successive substitutions discussed in Chapter 2, Section 7, for nonlinear boundary value problems. To demonstrate its working in eigenproblems, we return to Eq. (6.139) but this time take the boundary conditions $\phi(0) = \phi(1) = 0$ that do not change the eigenvalues. For the purpose of successive substitutions, we write Eq. (6.139) in the form

$$\phi''^{(n+1)} + \lambda^{(n+1)}\phi^{(n+1)} \sin(\phi^{(n)})/\phi^{(n)} = 0 \qquad (6.143)$$

and start the iterative process with $\phi^{(1)} = \alpha \sin \pi x$, where α is the chosen amplitude. When $\phi^{(n)}$ is substituted into it, Eq. (6.143) becomes a linear eigenproblem for $\phi^{(n+1)}$, which is computed with an arbitrary amplitude. Therefore, before this last $\phi^{(n+1)}$ is resubstituted into Eq. (6.143), it must be *rescaled* to the chosen amplitude α.

Figure 6.16 shows the improvement of $\phi^{(n)}$ for $\alpha = 3$ when $n = 1$ and then when $n = 2$. No higher values of n are shown since they overlap the curve for $n = 2$. The eigenvalues successively computed in this way are $\lambda^{(2)} = 39.8$,

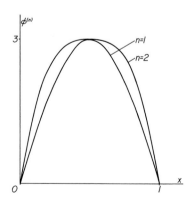

Fig. 6.16   Improvement of $\phi(x)$ in an axially compressed beam, using n successive substitutions according to Eq. (6.143).

$\lambda^{(3)} = 67.6$, $\lambda^{(4)} = 68.2$, $\lambda^{(5)} = 64.9$, $\lambda^{(6)} = 66.0$, $\lambda^{(7)} = 66.0$. But $\alpha = 3$ is a very strong nonlinear case.

EXERCISES

1. Use the beam element mass matrix in Eq. (6.138) and the beam element stiffness matrix in Eq. (4.53) to compute the lowest natural frequency of the cantilever beam, using different numbers of finite elements. Draw the convergence of the first eigenvalue versus the number of elements on a log–log scale and determine the rate of convergence.

2. Use the lumped element mass matrix

$$Me = \frac{h}{2} \begin{bmatrix} 1 & & & \\ & \alpha h^2 & & \\ & & 1 & \\ & & & \alpha h^2 \end{bmatrix}$$

to set up the algebraic eigenproblem for the cantilever beam. Employ seven elements in this discretization and study the dependence of the accuracy in the first five eigenvalues upon α.

3. Assemble the global mass and stiffness matrices in Exercise 1 from five finite elements. Then change the basis to condense the algebraic eigenproblem into a 3×3 system by taking as the new basis the finite element interpolates to x^2, x^3 and x^4. This you do by choosing the vectors z_1, z_2, and z_3 in Eq. (6.124), so as to have nodal values determined by x^2, x^3, and x^4, respectively. Compute the three eigenvalues from the condensed system and compare them with those computed from the full system.

4. The large displacement equation of equilibrium of a simply supported axially compressed column is

$$y'' + \lambda y(1 - y'^2)^{1/2} = 0, \qquad 0 < x < 1$$

$$y(0) = y(1) = 0$$

Use successive substitutions to solve this nonlinear eigenproblem and draw $y(\frac{1}{2})$ versus λ.

5. The linear equation of motion of the dangling chain (string) is

$$\frac{\partial^2 y}{\partial t^2} = g \frac{\partial}{\partial x}\left(x \frac{\partial y}{\partial x}\right), \qquad 0 < x < l, \qquad t > 0$$

where g is the gravitational acceleration. The x coordinate is vertical and the chain is hung at $x = l$ with the origin at the loose end of the chain. Separation of variables produces the eigenvalue problem

$$g(xy')' + \lambda y = 0, \qquad 0 < x < l$$

$$y(l) = 0, \qquad xyy' \to 0 \qquad \text{as} \quad x \to 0$$

Associate with this problem the energies

$$E(y,y) = \int_0^l xy'^2 \, dx, \qquad K(y,y) = \int_0^l y^2 \, dx$$

and show that the minimization of Rayleigh's quotient $R[y] = E(y,y)/K(y,y)$ with $y(1) = 0$ is equivalent to the eigenproblem. Use finite elements to compute the gravest natural frequency $\omega_1 = \sqrt{\lambda_1}$ of the chain and compare your result with the exact value

$$\omega_1 = \tfrac{1}{2}\sqrt{g/l}\,2.4048255577$$

In general,

$$\omega_n = \tfrac{1}{2}\sqrt{g/l}\,J_{0,n}$$

where $J_{0,n}$ is the nth zero of Bessel's function $J_0(x)$.

6. A beam is vertically clamped at its lower end and is left free at the top. Its equation of equilibrium is

$$py'''' + \rho g(xy')' = 0, \qquad 1 < x < l$$

where p includes the geometrical and elastic properties of the beam and ρ is the beam's mass per unit length. The x coordinate is vertical and has its origin at the free top end of the beam, which is otherwise clamped at $x = l$. Hence the essential conditions on the beam are

$$y(l) = y'(l) = 0$$

while at $x = 0$ the boundary conditions $y''(0) = y'''(0) = 0$ are natural. Associate with this problem the energies

$$E(y,y) = \int_0^l y''^2 \, dx, \qquad K(y,y) = \int_0^l xy'^2 \, dx$$

and verify that the minimization of Rayleigh's quotient produces the desired eigenproblem. Let $\omega^2 = \rho g/p$ and compute, using finite elements, the first ω, ω_1, at which the beam can assume also a drooped equilibrium position. Theoretically ω_1 is obtained from

$$J_{-1/3}(\tfrac{2}{3}\omega l^{3/2}) = 0 \qquad \text{and} \qquad \omega_1 = \tfrac{3}{2}l^{-3/2}1.8663$$

7. Let a clamped-free beam, as in Fig. 6.17, be acted upon by a tangent *follower* force on its free end. Assuming unit properties, the equation of

Fig. 6.17 Thin elastic cantilever beam with a follower force at the tip tangent to the middle line.

equation of motion of this beam is

$$y'''' + Py'' + \ddot{y} = 0, \qquad 0 < x < 1, \qquad t > 0$$

which the separation of variables $y(x, t) = y(x)e^{i\sqrt{\lambda}t}$ reduces to the eigenproblem

$$y'''' + Py'' - \lambda y = 0, \qquad 0 < x < 1$$

$$y(0) = y'(0) = y''(1) = y'''(1) = 0$$

Discretize this eigenvalue problem using finite differences and show that when $\lambda = 0$, there is no *static* bent equilibrium position to the beam. To study its *dynamic* equilibrium, increase P, starting from zero, and compute the fundamental natural frequency of the beam under load. Observe that at about $P = 20.19$ the motion of the beam *ceases to be periodic*, and it becomes dynamically unstable.

8. A clamped beam of length l rotates with an angular velocity Ω around an axis located at a distance R from the clamped support, as shown in Fig. 6.18. The fundamental frequencies ω of the beam are obtained from

$$\varepsilon \frac{d^4 y}{d\xi^4} - \frac{d}{d\xi}\left(q(\xi) \frac{dy}{d\xi} \right) = \left(\frac{\omega}{\Omega} \right)^2 y, \qquad 0 < \xi < 1$$

$$y(0) = y'(0) = 0, \qquad y''(1) = y'''(1) = 0$$

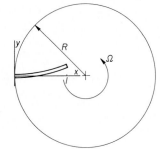

Fig. 6.18 A rotating cantilever beam.

where $\xi = x/l$ and

$$q(\xi) = (1 - \xi)[\tfrac{1}{2}(1 + \xi) - R/l]$$

Also

$$\varepsilon = p/\Omega^2 \rho a l^4$$

where p is the elastic modulus times the moment of inertia of the cross section, ρ the density, and a the area of its cross section. Assume $R/l = 1$ and proceed to compute $(\omega/\Omega)^2$ as a function of ε. When is $\omega_1 = 0$?

9. A uniform beam of length l is hinged at $x = 0$, being attached at this point to a spiral spring of constant k. It carries a mass m with rotary inertia j at $x = x_0$ and is otherwise free at $x = l$. Show that here

$$E(y, y) = \tfrac{1}{2} \int_0^l py''^2 \, dx + \tfrac{1}{2} ky'^2(0)$$

$$K(y, y) = \tfrac{1}{2} \int_0^l \rho a y^2 \, dx + \tfrac{1}{2} my^2(x_0) + \tfrac{1}{2} jy'^2(x_0)$$

where ρ denotes the density of the beam, a its cross section, and where y is subject only to the essential boundary condition $y(0) = 0$. Use one finite element to explore the dependence of the beam's fundamental natural frequency ω_1 upon the parameters of the problem. When $p = l = k = a = j = m = x_0 = 1$, $\omega_1 = 0.529$.

SUGGESTED FURTHER READING

Temple, G., and Bickley, W. G., *Rayleigh's Principle and its Applications to Engineering*. Constable Press, London, 1933.

Weinberger, H. F., *Variational Methods for Eigenvalue Problems*. Univ. of Minnesota Press, Minneapolis, 1962.

Collatz, L., *Eigenvalue Problems with Technical Applications*. Akademische Verlagsgesellschaft, Leipzig, 1963.

Panovko, Y. G., and Gubanova, I. I., *Stability and Oscillations of Elastic Systems*. Consultants Bureau, New York, 1965.

Gould, S. H., *Variational Methods for Eigenvalue Problems*. Univ. of Toronto Press, 1966.

Keller, J. B., and Antman, S., *Bifurcation Theory and Nonlinear Eigenvalue Problems*. Benjamin, New York, 1969.

Rubinstein, Moshe, F., *Structural Systems—Statics, Dynamics and Stability*. Prentice–Hall, Englewood Cliffs, New Jersey, 1970.

Weinstein, A., and Stenger, W., *Intermediate Problems for Eigenvalues*. Academic Press, New York, 1972.

Clough, R. W., and Penzien, J., *Dynamics of Structures*. McGraw-Hill, New York, 1975.

Leipholz, H., *Direct Variational Methods and Eigenvalue Problems in Engineering*. Noordhoff, Leyden, 1977.

7 Algebraic Properties of the Global Matrices

1. Eigenvalue Range in $Ky = \lambda My$

The algebraic eigenproblem created by either finite differences or finite elements is of the general form $Ky = \lambda My$, and in our forthcoming discussion on time dependent problems, in the next three chapters, we shall be pressed to have the range in which λ is found. We make it our first task then, in this section, to locate the discrete finite element spectrum, and we do it in terms of the computationally accessible eigenvalues of the *element* matrices k_e and m_e.

In fact, *if*

$$\omega \leqslant y_e{}^T k_e y_e / y_e{}^T m_e y_e \leqslant \Omega, \, e = 1, 2, \ldots, Ne \tag{7.1}$$

then also

$$\omega \leqslant y^T K y / y^T M y \leqslant \Omega \tag{7.2}$$

where k_e and m_e are the element matrices from which the global K and M matrices are assembled, respectively.

To prove the theorem in Eqs. (7.1) and (7.2), we recall that k_e and m_e are symmetric, with k_e usually positive semidefinite and m_e, by virtue of the linear independence of the shape functions and when $\rho > 0$, always positive definite. Rayleigh's theorem in its matrix version holds for these element matrices and for any typical element with matrices of dimension $n \times n$

$$\lambda_1{}^e y_e{}^T m_e y_e \leqslant y_e{}^T k_e y_e \leqslant \lambda_n{}^e y_e{}^T m_e y_e \tag{7.3}$$

125

where λ_1^e and λ_n^e are the lowest (1st) and highest (nth) eigenvalues of the element eigenproblem

$$k_e y_e = \lambda^e m_e y_e \tag{7.4}$$

Assembly of the k_e into K, together with Eq. (7.3), yields

$$y^T K y = \sum_{e=1}^{Ne} y_e^T k_e y_e \leq \max_e \{\lambda_n^e\} \sum_e y_e^T m_e y_e$$
$$\leq \max_e \{\lambda_n^e\} y^T M y \tag{7.5}$$

Also in precisely the same way,

$$y^T K y \geq \min_e \{\lambda_1^e\} \sum_{e=1}^{Ne} y_e^T m_e y_e \geq \min_e \{\lambda_1^e\} y^T M y \tag{7.6}$$

Our sought bound follows readily from Eqs. (7.5) and (7.6) in the form

$$\min_e \{\lambda_1^e\} \leq y^T K y / y^T M y \leq \max_e \{\lambda_n^e\} \tag{7.7}$$

where e ranges over all Ne finite elements.

As a simple example for Eq. (7.7), consider the eigenproblem $-y'' = \lambda y$, $y(0) = y(1) = 0$ discretized with the first-order string elements

$$k_e = \frac{1}{h} \begin{bmatrix} 1 & -1 \\ -1 & 1 \end{bmatrix}, \qquad m_e = \frac{h}{6} \begin{bmatrix} 2 & 1 \\ 1 & 2 \end{bmatrix} \tag{7.8}$$

Quick computations show that here

$$0 \leq y_e^T k_e y_e / y_e^T m_e y_e \leq 12h^{-2} \tag{7.9}$$

and hence, by Eq. (7.7), also

$$0 \leq y^T K y / y^T M y \leq 12h^{-2} \tag{7.10}$$

Because k_e in Eq. (7.8) is only positive *semidefinite*, Eq. (7.7) fails to ascertain that actually $y^T K y / y^T M y > 0$. On the other hand, the *upper bound*, the one that is the most expedient in time-dependent problems, is established in Eq. (7.10) realistically.

With the lumped element mass matrix

$$m_e = \frac{1}{2} h \begin{bmatrix} 1 & \\ & 1 \end{bmatrix} \tag{7.11}$$

Eq. (7.7) predicts that

$$0 < y^T K y / y^T M y < 4h^{-2} \tag{7.12}$$

To check the effectiveness of the upper bound in Eq. (7.12), we observe that the global system assembled from k_e in Eq. (7.8) and m_e in Eq. (7.11) is in fact that in Eq. (6.35), obtained with finite differences. A closed form expression for the eigenvalues of this system is given in Eq. (6.46), where we have that $\lambda_N = 4h^{-2}\sin^2[\pi N/2(N+1)]$, and as $N \to \infty$, $\lambda_N \to 4h^{-2}$. The upper bound in Eq. (7.12) is thus sharp. Notice that in this case, Gerschgorin's theorem produces the same bounds as in Eq. (7.12), but when M is nondiagonal, Eq. (7.7) is manifestly preferable to Gerschgorin's theorem.

As a further example of the use of Eq. (7.7), consider the general Sturm–Liouville second-order eigenproblem $-(p(x)y')' + q(x)y = \lambda\rho(x)y$, discretized with linear finite elements. The corresponding element eigenproblem is here

$$\left\{\frac{1}{h}p_c\begin{bmatrix} 1 & -1 \\ -1 & 1 \end{bmatrix} + \frac{1}{2}hq_c\begin{bmatrix} 1 & \\ & 1 \end{bmatrix}\right\}y_e = \lambda^e\left\{\frac{1}{2}h\rho_c\begin{bmatrix} 1 & \\ & 1 \end{bmatrix}\right\}y_e \qquad (7.13)$$

where p_c, q_c, and ρ_c are the values of $p(x)$, $q(x)$, and $\rho(x)$, respectively, at the center of the element, and where numerical integration with lumping was used in the formation of the element matrices. Presently

$$\min\left\{\frac{q_c}{\rho_c}\right\} \leqslant y^T K y/y^T M y \leqslant \max\left\{4\frac{p_c}{\rho_c}h^{-2} + \frac{q_c}{\rho_c}\right\} \qquad (7.14)$$

A more striking example, showing the superiority of Eq. (7.7) over Gerschgorin's theorem, is provided by the beam finite elements

$$k_e = \frac{1}{h^3}\begin{bmatrix} 12 & 6 & -12 & 6 \\ 6 & 4 & -6 & 2 \\ -12 & -6 & 12 & -6 \\ 6 & 2 & -6 & 4 \end{bmatrix}, \qquad m_e = \frac{h}{420}\begin{bmatrix} 156 & 22 & 54 & -13 \\ 22 & 4 & 13 & -3 \\ 54 & 13 & 156 & -22 \\ -13 & -3 & -22 & 4 \end{bmatrix}$$

$$(7.15)$$

Gerschgorin's theorem fails even to determine the positive definiteness of m_e, while Eq. (7.7) assures us that

$$0 \leqslant y^T K y/y^T M y \leqslant 8460h^{-4} \qquad (7.16)$$

In all the above examples, we disregarded the essential boundary conditions that we remember are enforced by deflating the element and later the global matrices. But these modifications in k_e and m_e only *raise* λ_1^e and *lower* λ_n^e, and can therefore be conveniently ignored.

2. Spectral Norms of K and M

While in the previous section we established bounds on the eigenvalues of $Ky = \lambda My$, in this section we endeavor to bound the eigenvalues of K and M themselves. We shall seek, in particular, a nontrivial lower bound on the smallest eigenvalue of K in order to establish properly the positive definiteness of the global stiffness matrix.

To do this, we first need the inequality

$$y^T y \leqslant \sum_{e=1}^{Ne} y_e^T y_e \leqslant y^T y \eta_{max} \tag{7.17}$$

in which η_{max} is the maximal number of elements meeting at a nodal point. The veracity of Eq. (7.17) should be obvious. Let it be illustrated by the finite element arrangements in Fig. 7.1a,b. For the first, $y^T = (y_1, y_2, y_3, y_4, y_5, y_6, y_7)$, while $y_1^T = (y_1, y_2, y_3)$, $y_2^T = (y_3, y_4, y_5)$, $y_3^T = (y_5, y_6, y_7)$, and

$$y^T y \leqslant \sum_{e=1}^{3} y_e^T y_e = y_1^2 + y_2^2 + 2y_3^2 + 2y_4^2 + 2y_5^2 + y_6^2 + y_7^2 \leq 2y^T y \tag{7.18}$$

For the second mesh (see Fig. 7.1b) $y^T = (y_1, y_1', y_2, y_2', y_3, y_3')$, while $y_1^T = (y_1, y_1', y_2, y_2')$, $y_2^T = (y_2, y_2', y_3, y_3')$, and

$$y^T y \leqslant \sum_{e=1}^{2} y_e^T y_e \leqslant y_1^2 + y_1'^2 + 2(y_2^2 + y_2'^2) + y_3^2 + y_3'^2 \leqslant 2y^T y \tag{7.19}$$

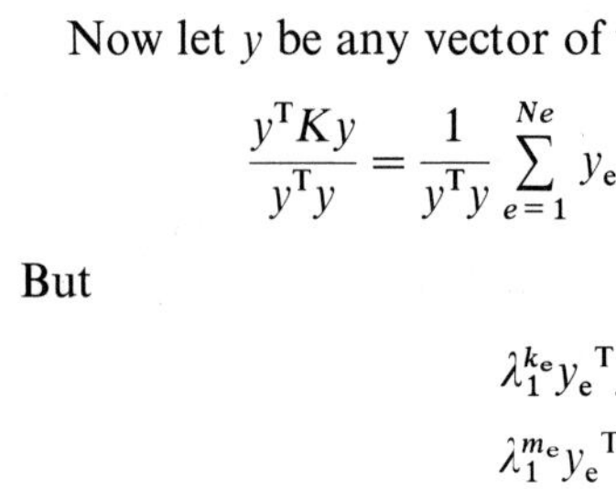

Fig. 7.1 (a) Three quadratic C^0 elements, and (b) two cubic C^1 elements.

Now let y be any vector of the proper dimension; then by definition,

$$\frac{y^T K y}{y^T y} = \frac{1}{y^T y} \sum_{e=1}^{Ne} y_e^T k_e y_e, \qquad \frac{y^T M y}{y^T y} = \frac{1}{y^T y} \sum_{e=1}^{Ne} y_e^T m_e y_e \tag{7.20}$$

But

$$\lambda_1^{k_e} y_e^T y_e \leqslant y_e^T k_e y_e \leqslant \lambda_n^{k_e} y_e^T y_e$$
$$\lambda_1^{m_e} y_e^T y_e \leqslant y_e^T m_e y_e \leqslant \lambda_n^{m_e} y_e^T y_e \tag{7.21}$$

where $\lambda_1^{k_e}$, $\lambda_n^{k_e}$, $\lambda_1^{m_e}$ and $\lambda_n^{m_e}$ are the lowest (1st) and highest (nth) eigenvalues of the element matrices k_e and m_e, and with Eqs. (7.17) and (7.20), Eq. (7.21) produces the bounds

$$\min_{e}\{\lambda_1^{k_e}\} \leqslant y^T K y / y^T y \leqslant \max_{e}\{\lambda_n^{k_e}\}\eta_{max}$$

$$\min_{e}\{\lambda_1^{m_e}\} \leqslant y^T M y / y^T y \leqslant \max_{e}\{\lambda_n^{m_e}\}\eta_{max} \tag{7.22}$$

The second inequality in Eq. (7.22) duly establishes the positive definiteness of M, while that in the first only confirms that K is positive semidefinite if any one of the element stiffness matrices k_e is singular, as it usually is. For instance, when K is assembled from

$$k_e = \frac{1}{h}\, p_c \begin{bmatrix} 1 & -1 \\ -1 & 1 \end{bmatrix} + \frac{1}{2}\, h q_c \begin{bmatrix} 1 & \\ & 1 \end{bmatrix} \tag{7.23}$$

Eq. (7.22) predicts that

$$\min\{\tfrac{1}{2}hq_c\} \leqslant y^T K y / y^T y \leqslant \max\{(4/h)p_c + hq_c\} \tag{7.24}$$

and if $q_c = 0$ the lower bounds fail, the same way Gerschgorin's theorem does, to establish the positive definiteness of K. On the other hand, the same bound for M never fails. Consider, for example, the element mass matrix $(420/h)m_e$ in Eq. (4.47). Its eigenvalues are $\lambda_1^{m_e} = \frac{1}{5}h^2$, $\lambda_2^{m_e} = \frac{7}{6}h^2$, $\lambda_3^{m_e} = 102 + \frac{9}{8}h^2$, and $\lambda_4^{m_e} = 210 + 6h^2$, and hence for M assembled from this element mass matrix we have that

$$\tfrac{1}{5}h^2 \leqslant y^T M y / y^T y \leqslant 420 + 12h^2 \tag{7.25}$$

What makes the bounds in Eq. (7.22) so attractive is that they explicitly involve only the discretization parameters. No elusive intrinsic information about the boundary value problem is needed in them, and their *numerical* realization is therefore immediate. Yet they are not sophisticated enough to produce a lower nonzero bound on the first eigenvalue of K. In order to produce a realistic lower bound on the spectrum of K, we are obliged to include in the bounding theorem a characteristic number of the relevant differential operator. Let λ_1, the lowest eigenvalue of the differential operator under consideration, be this characteristic number, and $\hat{\lambda}_1$ the corresponding finite element eigenvalue computed from $Ky = \lambda My$. Then,

The lowest eigenvalue $\lambda_1{}^K$ of the finite element global stiffness matrix K is bounded by

$$\lambda_1 \min_{e}\{\lambda_1^{m_e}\} \leqslant \lambda_1{}^K \leqslant \hat{\lambda}_1 \max_{e}\{\lambda_n^{m_e}\}\eta_{max} \tag{7.26}$$

where η_{max} is the maximum number of elements meeting at any nodal point.

Rayleigh's principle, stating that $y^T K y / y^T M y \geqslant \lambda_1$, serves as the basis upon which the proof of Eq. (7.26) is founded. For with it, we have that

$$y^T K y / y^T y \geqslant \lambda_1 y^T M y / y^T y \tag{7.27}$$

and if we make the particular choice of y to be the eigenvector corresponding to $\lambda_1{}^K$, then Eq. (7.27) becomes

$$\lambda_1{}^K \geqslant \lambda_1 y^T M y / y^T y \tag{7.28}$$

with which Eq. (7.22) turns into

$$\lambda_1{}^K \geqslant \lambda_1 \min_e \{\lambda_1^{m_e}\} \tag{7.29}$$

To get a similar upper bound on $\lambda_1{}^K$, we assume y to be the eigenvector corresponding to $\hat{\lambda}_1$, such that

$$y^T K y / y^T y = \hat{\lambda}_1 y^T M y / y^T y \tag{7.30}$$

But $y^T K y / y^T y \geqslant \lambda_1{}^K$ and consequently

$$\lambda_1{}^K \leqslant \hat{\lambda}_1 y^T M y / y^T y \tag{7.31}$$

from which, using Eq. (7.22), the right-hand side bound on $\lambda_1{}^K$ in Eq. (7.26) readily follows.

Equation (7.22) furnishes useful upper bounds on $\lambda_N{}^K$ and $\lambda_N{}^M$—the maximal (Nth) eigenvalues of K and M. To derive a lower bound on these eigenvalues, we select y such that a portion of it is the eigenvector corresponding to $\max_e\{\lambda_n^{k_e}\}$ or $\max_e\{\lambda_n^{m_e}\}$ and the rest zero, and have from this that

$$\lambda_N{}^K \geqslant \max_e \{\lambda_n^{k_e}\}, \qquad \lambda_N{}^M \geqslant \max_e \{\lambda_n^{m_e}\} \tag{7.32}$$

Combined with the bounds in Eq. (7.22), this yields

$$\max_e \{\lambda_n^{k_e}\} \leqslant \lambda_N{}^K \leqslant \max_e \{\lambda_n^{k_e}\} \eta_{\max}$$

$$\max_e \{\lambda_n^{m_e}\} \leqslant \lambda_N{}^M \leqslant \max_e \{\lambda_n^{m_e}\} \eta_{\max} \tag{7.33}$$

As an example of the bounds in Eqs. (7.22), (7.26), and (7.33), consider the case of a hanging chain discretized with linear finite elements for which

$$k_e = \frac{1}{2}(2e - 1)\begin{bmatrix} 1 & -1 \\ -1 & 1 \end{bmatrix}, \qquad m_e = \frac{h}{6}\begin{bmatrix} 2 & 1 \\ 1 & 2 \end{bmatrix} \tag{7.34}$$

If the chain is of unit length and free to oscillate in a unit gravitational field, then it is known that $\lambda_1 = 1.45$, while from Eq. (7.34), we have that $\lambda_1^{k_e} = 0$,

$\lambda_2^{k_e} = 2/h$, $\lambda_1^{m_e} = \frac{1}{6}h$, and $\lambda_2^{m_e} = \frac{1}{2}h$. Substitution of these values into Eqs. (7.22), (7.26), and (7.33) furnishes us with the bounds

$$\frac{1}{6}h \leqslant y^T M y / y^T y \leqslant h \qquad (7.35)$$

and

$$0.26h \leqslant \lambda_1^{K} \leqslant 1.45h \qquad (7.36)$$

assuming that $\lambda_1 = \hat{\lambda}_1$; and also

$$2h^{-1} \leqslant \lambda_N^{K} \leqslant 4h^{-1} \qquad (7.37)$$

3. Spectral Condition Numbers

The spectral or l_2 norms of the positive definite and symmetric *stiffness* matrix K and *flexibility* matrix K^{-1} are

$$\|K\|_2 = \lambda_N^{K}, \qquad \|K^{-1}\|_2 = 1/\lambda_1^{K} \qquad (7.38)$$

and the spectral condition number $C_2(K) = \|K\|_2 \|K^{-1}\|_2$ of K is therefore

$$C_2(K) = \lambda_N^{K}/\lambda_1^{K} \qquad (7.39)$$

This important number, so crucial in the perturbation analysis of linear algebraic systems, is bound according to the analysis of the previous section by

$$\frac{1}{\hat{\lambda}_1} \frac{\max_e\{\lambda_n^{k_e}\}}{\max_e\{\lambda_n^{m_e}\}} \frac{1}{\eta_{max}} \leqslant C_2(K) \leqslant \frac{1}{\lambda_1} \frac{\max_e\{\lambda_n^{k_e}\}}{\min_e\{\lambda_1^{m_e}\}} \eta_{max} \qquad (7.40)$$

as can be verified by the introduction of Eqs. (7.26) and (7.33) into the definition of $C_2(K)$ in Eq. (7.39). For the global mass matrix M, we have from Eqs. (7.22) and (7.33) that

$$1 \leqslant C_2(M) \leqslant (\max_e\{\lambda_n^{m_e}\}/\min_e\{\lambda_1^{m_e}\})\eta_{max} \qquad (7.41)$$

When the mesh of finite elements is fine enough, $\hat{\lambda}_1$ may be sufficiently accurate for λ_1 to be substituted for it. Also, even though the derivation of the upper bound on $C_2(K)$ in Eq. (7.40) was based on strict adherence to the conditions of Rayleigh's principle, this bound is equally useful when numerical integration or lumping is involved except that now λ_1 must be understood to be within the approximation accuracy.

Equations (7.40) and (7.41) are used next to study the behavior of the spectral condition numbers of K and M set up with various common finite elements. First we look at the discretization of $-y'' = f(x)$, $0 < x < 1$, $y(0) = y(1) = 0$ with linear first-order finite elements. Here $\lambda_1^{me} = \frac{1}{6}h$, $\lambda_2^{ke} = 2/h$, $\lambda_2^{me} = \frac{1}{2}h$, $\lambda_1 = \pi^2$, and we have from Eqs. (7.40) and (7.41) that

$$(2/\pi^2)h^{-2} \leqslant C_2(K) \leqslant (24/\pi^2)h^{-2} \tag{7.42}$$

and

$$1 \leqslant C_2(M) \leqslant 6 \tag{7.43}$$

In the case in which the same problem is discretized with quadratic finite elements, whose element matrices are given in Eq. (4.42), then, since here $\lambda_1^{me} = \frac{4}{25}h$, $\lambda_3^{me} = \frac{16.6}{15}h$, $\lambda_3^{ke} = 4/h$, and $\lambda_1 = \pi^2$, we get from Eqs. (7.26) and (7.33) that

$$4/h \leqslant \lambda_N^K \leqslant 8/h \tag{7.44}$$

and

$$0.16\pi^2 h \leqslant \lambda_1^K \leqslant 2.2\pi^2 h \tag{7.45}$$

and consequently

$$(1.82/\pi^2)h^{-2} \leqslant C_2(K) \leqslant (50/\pi^2)h^{-2} \tag{7.46}$$

as compared with the computed value $C_2(K) = (8/\pi^2)h^{-2}$.

An improved upper bound on $C_2(K)$ can often be secured by using lumping. Take the last quadratic element, we may write its element mass matrix as

$$m_e = \frac{2}{3}h \begin{bmatrix} 1 & & \\ & 1 & \\ & & 1 \end{bmatrix} \tag{7.47}$$

which, even though it does not assure the optimal rate of convergence for the eigenvalues, does lead nevertheless to the slower covergence of λ_1. When more than two elements are used in the discretization, $\hat{\lambda}_1$ becomes for all practical purposes interchangeable with λ_1, and Eq. (7.40) may still be used with the lumped m_e given in Eq. (7.47). But with this m_e, we obtain the better bound

$$C_2(K) \leqslant (12/\pi^2)h^{-2} \tag{7.48}$$

New insight into the ways in which Eq. (7.40) can be applied to bound $C_2(K)$ is provided by the fixed circular membrane. The relevant energy norm here is

$$\|y\|_1^2 = \int_0^1 ry'^2 \, dr \tag{7.49}$$

and for a first-order element extending over $r_1 \leqslant r \leqslant r_2$, Eq. (7.49) delivers

$$k_e = \frac{1}{h} \frac{r_1 + r_2}{2} \begin{bmatrix} 1 & -1 \\ -1 & 1 \end{bmatrix} \tag{7.50}$$

If we choose to compute λ_1, the fundamental eigenvalue of the membrane, from

$$-\frac{1}{r} \frac{d}{dr} \left(r \frac{dy}{dr} \right) = \lambda y, \qquad 0 < r < 1 \tag{7.51}$$

which assumes constant unit density distribution in the membrane, then

$$K(y, y) = \int_0^1 r y^2 \, dr \tag{7.52}$$

and with some approximations

$$m_e = \frac{1}{2} (r_1 + r_2) h \begin{bmatrix} 1 & \\ & 1 \end{bmatrix} \tag{7.53}$$

Now $\min_e \{\lambda_1^{m_e}\} = \frac{1}{2} h^2$, $\max_e \{\lambda_2^{m_e}\} = h$, $\max_e \{\lambda_2^{k_e}\} = 2/h$, and we know that for this circular membrane, $\lambda_1 = (2.404826)^2$. Substituting all that into Eq. (7.40), we get

$$(1/\lambda_1) h^{-2} \leqslant C_2(K) \leqslant (8/\lambda_1) h^{-3} \tag{7.54}$$

and the bounds diverge as $h \to 0$.

To correct this spread, we propose to compute λ_1 from a *variable* density distribution and write Eq. (7.51) as

$$-\frac{d}{dr} \left(r \frac{dy}{dr} \right) = \lambda r \rho y, \qquad 0 < r < 1 \tag{7.55}$$

$$y(1) = 0$$

and choose $r\rho = 1$. In so doing, we secured an element mass matrix, equal for all elements, of the form

$$m_e = \frac{h}{2} \begin{bmatrix} 1 & \\ & 1 \end{bmatrix} \tag{7.56}$$

and $\lambda_1^{m_e} = \lambda_2^{m_e} = h/2$. Equation (7.55) with $\rho r = 1$ is recognized to be the eigenvalue problem of the hanging chain, for which we know that $\lambda_1 = (2.404826/2)^2$. Hence, now for the same global stiffness matrix K as in Eq. (7.54), we have the better bounds

$$(2/\lambda_1) h^{-2} \leqslant C_2(K) \leqslant (8/\lambda_1) h^{-2} \tag{7.57}$$

One of the most remarkable conclusions reached from the above discussion is that in second-order problems $C_2(K) = O(h^{-2}) = O(Ne^2)$ *independently of the degree of the shape functions.* It also informs us that the refinement of the mesh, intended to improve the discretization accuracy, invariably causes the condition of the global stiffness matrix K to decline. On the other hand, we are assured by the above analysis that for a reasonable number of elements in the discretization, the condition number of K is far below the level that may cause numerical singularity. On a computer holding six decimals, the numerical existence of K^{-1} becomes doubtful when $C_2(K) = 10^6$. With quadratic string elements, this dangerous ill condition is reached only when $h = 10^{-3}$, or with two thousand elements!

Application of Eq. (7.40) to fourth-order problems with C^1 elements is as straightforward as it was for the previous second-order problem. Consider the uniform elastic beam discretized with cubic elements, each with the nodal values y_1, hy_1', y_2, hy_2', whose element matrices are given in Eq. (7.15). Here $\lambda_1^{me} = \frac{1}{180}h$, $\lambda_4^{me} = \frac{1}{6}h$, $\lambda_4^{ke} = 12\sqrt{5}/h^3$ and if the beam is assumed to be simply supported [i.e., $y(0) = y(1) = y''(0) = y''(1) = 0$], $\lambda_1 = \pi^4$. Assuming further that $\lambda_1 = \hat{\lambda}_1$, we obtain from Eq. (7.40) the bounds

$$h^{-4} \leqslant C_2(K) \leqslant 99h^{-4} \tag{7.58}$$

Numerically, we find that for this beam $C_2(K) = 0.49\ Ne^4$, $Ne > 1$.

Far enough in the discretization when there is a large number of elements, or when h is small, a closer bound on $C_2(K)$ can be found through lumping as follows: We write the element mass matrix m_e for the cubic beam in the diagonal form

$$m_e = \frac{h}{2}\begin{bmatrix} 1 & & & \\ & \alpha h^2 & & \\ & & 1 & \\ & & & \alpha h^2 \end{bmatrix} \tag{7.59}$$

which assures a correct kinetic energy for a constant y. A correct kinetic energy for the linear y is obtained from m_e in Eq. (7.59) with the *negative* $\alpha = -\frac{1}{6}$. But a negative or even a zero entry in this diagonal m_e does not allow the use of Eq. (7.40), and we therefore forfeit the conservation of kinetic energy for the linear y and choose $\alpha = 1$. This makes a crude approximation for m_e, so much so that discretization of a simply supported beam with 4, 6, 8, 10, 12, and 14 elements delivers only 59.8, 76.4, 84.3, 88.7, 91.6, and 93.3, respectively, for $\lambda_1 = 97.41$. Nevertheless, with sufficiently many elements, λ_1 may still be replaced with $\hat{\lambda}_1$ and we may use this lumped m_e in Eq. (7.40). With $h = 1$ inside the element mass matrix of Eq. (7.59), we now get the lower upper bound

$$C_2(K) \leqslant 1.1h^{-4}, \qquad h \ll 1 \qquad (7.60)$$

It is not difficult to show, using Eq. (7.40), that in general, beam elements of any order have a $C_2(K) = O(h^{-4})$. This faster growing ill condition of the global stiffness matrix K poses a considerably larger threat to the successful solution of the discrete equilibrium equation than it did for the string.

4. Irregular Meshes

All the bounds derived in the previous sections for $\lambda_1{}^K$, $\lambda_1{}^M$, $\lambda_N{}^K$, and $\lambda_N{}^M$ are general and apply to uniform as well as to nonuniform meshes of finite elements. But the bounds on $\lambda_1{}^K$ in Eq. (7.26) part when the elements in the mesh are of widely different sizes. Responsible for the widening difference between these bounds is the element mass matrix m_e, whose eigenvalues are proportional to the element size. We already encountered the difficulty of providing close bounds on $C_2(K)$ when the eigenvalues of m_e vary greatly with e in the circular membrane, and solved it with a variable density distribution in the eigenvalue problem from which λ_1 is computed. The same can be done here too. The element mass matrix is formed from

$$K(y, y) = \int_0^1 \rho(x) y^2 \, dx \qquad (7.61)$$

$\rho(x)$ is adjusted so that $\min_e\{\lambda_1^{m_e}\}/\max_e\{\lambda_n^{m_e}\}$ is as close to one as possible, and λ_1 is computed with this nonuniform distribution of ρ. An obvious restriction on $\rho(x)$ is that it not cause a zero λ_1. Also, in this technique, λ_1 becomes dependent on the mesh parameters and it will be our added task to bound it too.

To fix ideas, let us analyze the condition of K for $y'' + f(x) = 0$, $y(0) = y(1) = 0$, formed with a nonuniform mesh of linear finite elements for which

$$k_e = \frac{1}{h_e}\begin{bmatrix} 1 & -1 \\ -1 & 1 \end{bmatrix}, \qquad h_{min} \leqslant h_e \leqslant h_{max} \qquad (7.62)$$

First we assume $\rho(x) = 1$, compute λ_1 from $-y'' = \lambda y$, $y(0) = y(1) = 0$, and hence $\lambda_1^{m_e}$ from

$$m_e = \frac{1}{6}h_e\begin{bmatrix} 2 & 1 \\ 1 & 2 \end{bmatrix}, \qquad h_{min} \leqslant h_e \leqslant h_{max} \qquad (7.63)$$

and have from this and Eq. (7.26) that

$$\tfrac{1}{6}\pi^2 h_{min} \leqslant \lambda_1{}^K \leqslant \pi^2 h_{max} \tag{7.64}$$

Equation (7.33) provides us with

$$2/h_{min} \leqslant \lambda_N{}^K \leqslant 4/h_{min} \tag{7.65}$$

and consequently

$$(2/\pi^2)h_{min}^{-1}h_{max}^{-1} \leqslant C_2(K) \leqslant (24/\pi^2)h_{min}^{-2} \tag{7.66}$$

To tighten the bounds on $C_2(K)$, we compute λ_1 from $y'' + \lambda\rho(x)y = 0$, $y(0) = y(1) = 0$, and find it simplest to assume $\rho(x)$ constant within each element. Then

$$m_e = \frac{1}{6}\rho_e h_e \begin{bmatrix} 2 & 1 \\ 1 & 2 \end{bmatrix} \tag{7.67}$$

and to keep $\lambda_1^{m_e}$ and $\lambda_2^{m_e}$ independent of h_e, we choose

$$h_1\rho_1 = h_2\rho_2 = h_3\rho_3 = \cdots \tag{7.68}$$

We also restrict ourselves to unit total mass or

$$\int_0^1 \rho(x)\,dx = 1 \tag{7.69}$$

and conclude from this that

$$h_e\rho_e = 1/Ne \tag{7.70}$$

where Ne is the total number of elements. Now

$$m_e = \frac{1}{6N_e}\begin{bmatrix} 2 & 1 \\ 1 & 2 \end{bmatrix} \tag{7.71}$$

from which $\lambda_1^{m_e} = 1/(6Ne)$ and $\lambda_2^{m_e} = 1/(2Ne)$ are computed. With these values, it occurs that

$$\tfrac{1}{6}\lambda_1 Ne^{-1} \leqslant \lambda_1{}^K \leqslant \lambda_1 Ne^{-1} \tag{7.72}$$

and then

$$C_2(K) \leqslant (24/\lambda_1)Neh_{min}^{-1} \tag{7.73}$$

But Eq. (7.73) is not useful yet because λ_1 is still a function of the discretization parameters, as a result of the particular density distribution $\rho_e = 1/(h_e Ne)$. In what follows we shall bound λ_1 for any unit mass arbitrarily distributed.

Fig. 7.2   A beam and a string with all their mass at the center point.

beam   *string*

To arrive at this bound, let y be the eigenfunction corresponding to λ_1 such that

$$\lambda_1 = E(y, y)/K(y, y) \tag{7.74}$$

But

$$K(y, y) = \int_0^1 \rho(x)y^2 \, dx \leqslant \max_x\{y^2\} \int_0^1 \rho(x) \, dx \tag{7.75}$$

and since the total mass equals one, as expressed in Eq. (7.69), we have that the characteristic number Φ defined by

$$\Phi = \min_y\{E(y, y)/\max_x\{y^2\}\} \tag{7.76}$$

is such that

$$\Phi \leqslant \lambda_1 \tag{7.77}$$

for any density distribution $\rho(x)$, provided that the total mass is constant. According to Eq. (7.76), Φ is the fundamental eigenvalue of the elastic body when all the mass is concentrated at the crest of the corresponding eigenfunction. To compute Φ, the mass is assumed at a point which is varied in location until the lowest fundamental eigenvalue is reached. When the body is symmetric and with symmetric boundary conditions, Φ occurs when all the mass is piled at the center point as in Fig. 7.2. For a string that is fixed at both its ends and a beam that is simply supported to the right and to the left, $\Phi = 4$ and $\Phi = 48$, respectively, as compared with $\lambda_1 = \pi^2$ for the string and $\lambda_1 = \pi^4$ for the beam when $\rho = 1$. With $\lambda_1 \geqslant \Phi = 4$, Eq. (7.76) becomes

$$C_2(K) \leqslant 6Neh_{min}^{-1} \tag{7.78}$$

which is better than the bound in Eq. (7.66) when $h_{min} \leqslant 4/(\pi^2 Ne)$.

Both the string and the beam are capable of carrying point inertia forces, as is the (unit) circular plate for which $\Phi = 16\pi$, in the sense that the concentration of their masses at one point does not cause the trivial bound $\lambda_1 \geqslant \Phi = 0$. It is not the same for the circular membrane that, having a point mass, has a zero fundamental natural frequency. So, instead of concentrating the mass at a single point, we collect it all inside a ring, having the width of the smallest element and displace this interval to minimize λ_1. In the case of a circular membrane, its frequency is lowest when the mass is put nearest to the center in the form

$$\rho(r) = \rho, \quad 0 \leqslant r \leqslant h; \quad \rho(r) = 0, \quad h < r \leqslant 1 \tag{7.79}$$

The eigenvalue problem we need to solve now is

$$-\frac{1}{r}(ry')' = \lambda \rho(r)y, \qquad 0 < r < 1$$

$$y(1) = 0, \qquad y'(0) = 0$$

(7.80)

and in the interval $h < r \leqslant 1$ where $\rho(r) = 0$

$$y = c_1 \log(1/r)$$

(7.81)

while in the interval $0 \leqslant r \leqslant h$, where $\rho(r)$ is the constant ρ, we assume y to be approximately given by

$$y = c_2 + c_3 r^2$$

(7.82)

and get upon the substitution of y in Eq. (7.82) into the eigenproblem (7.80) that

$$4c_3 + \lambda_1 \rho c_2 + \lambda_1 \rho c_2 r^2 = 0$$

(7.83)

Neglecting r^2 yields us

$$c_3 = -\tfrac{1}{4}\lambda_1 \rho c_2$$

(7.84)

such that in the range $0 \leqslant r \leqslant h$

$$y = c_2(1 - \tfrac{1}{4}\lambda_1 \rho r^2)$$

(7.85)

To fix λ_1, we match y and y' from Eqs. (7.81) and (7.85) at $r = h$ and have, with $\rho h^2/2 = 1$, that

$$\lambda_1 = \frac{2}{1 + 2\log(1/h)}$$

(7.86)

or

$$\Phi = \frac{2}{1 + 2\log(1/h_{min})}$$

(7.87)

Because $\log(1/h)$ is such a slowly growing function, Φ is nearly independent of the mesh. Between $h_{min} = \tfrac{1}{2}$ and $h_{min} = \tfrac{1}{100}$, Φ changes only between 0.8 and 0.2.

5.   The Influence (Green's) Function

The nodal response to a unit force acting at a node is the corresponding column in the flexibility matrix K^{-1}. This is the discrete counterpart to the

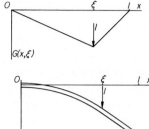

Fig. 7.3   Green's function $G(x, \xi)$ for the fixed string.

Fig. 7.4   Green's function $G(x, \xi)$ for the cantilever beam.

influence (or Green's) function, which plays an important role in our subsequent investigations of the properties of the global stiffness matrix K and its inverse K^{-1}. Green's function $G(x, \xi)$ is the response at x due to the application of point unit force at ξ. Figure 7.3 shows Green's function for the fixed string. At point ξ, a unit force acts, causing the deflection $G(x, \xi)$ where

$$G(x, \xi) = \begin{cases} x(1 - \xi), & \xi \geq x \\ \xi(1 - x), & \xi \leq x \end{cases} \tag{7.88}$$

Green's function for the cantilever beam shown in Fig. 7.4 is found to be

$$G(x, \xi) = \begin{cases} \frac{1}{2}\xi x^2 - \frac{1}{6}x^3, & x \leq \xi \\ -\frac{1}{6}\xi^3 + \frac{1}{2}\xi^2 x, & x \geq \xi \end{cases} \tag{7.89}$$

and we observe that for the string as well as for the beam Green's function is *symmetric*, or $G(x, \xi) = G(\xi, x)$. Also, when $x = \xi$, $G(x, x)$ represents the deflection of either string or beam at the point of the force action, and in both these cases $G(x, x)$ is positive and finite. The influence function is also finite for the circular plate, but is infinite for the circular membrane.

The symmetry of Green's function depends on differential equation and boundary conditions. Boundary value problems for which Green's function is symmetric are termed symmetric or *self-adjoint*. These problems are associated with a quadratic energy expression $E(y, y)$ that generates symmetric global finite element stiffness matrices. Upon its inversion, the symmetric global stiffness matrix K produces a symmetric flexibility matrix K^{-1} which is the discrete counterpart to Green's function, the same way K is the discrete counterpart to the differential operator. Symmetry or self-adjointness is an intrinsic property of the boundary value problem that is preserved in the finite element discretization.

A unit point torque acting on the string or beam consists of a pair of opposite point forces of the magnitude $P = 1/\varepsilon$ acting at a distance ε part, where $\varepsilon \to 0$. The displacement response at x to a unit torque at ξ is $\partial G(x, \xi)/\partial \xi$, and the slope response to the torque is consequently $\partial^2 G(x, \xi)/(\partial \xi\, \partial x)$. We verify from Eq. (7.88) that for the string a point torque causes a discontinuous displacement response, or an infinite slope response at ξ. For the beam it results from Eq. (7.89) that the displacement response to a point torque is C^1 and the slope response is finite.

A distributed force over x can be considered to be composed of point forces $f(\xi)\,d\xi$, with $G(\xi, x)f(\xi)\,d\xi$, the corresponding response at point x. Superposition of the responses at x from all points ξ expresses the total deflection $y(x)$ in the form

$$y(x) = \int_0^1 G(x, \xi)f(\xi)\,d\xi \tag{7.90}$$

In eigenproblems, $f(x)$ is replaced by $\lambda\rho(x)y$, by which Eq. (7.90) becomes

$$y(x) = \lambda \int_0^1 G(x, \xi)\rho(\xi)y(\xi)\,d\xi \tag{7.91}$$

This Green's function formulation of the eigenproblem can be used to establish a lower bound—the Φ number of the previous section—on λ_1. We briefly restrict ourselves to positive influence functions like those for the string and the beam in Eqs. (7.88) and (7.89) and proceed to show that, in this case, the eigenfunction y_1 corresponding to the fundamental eigenvalue λ_1 is also positive. To this end we multiply both sides of Eq. (7.91) by $y(x)\,dx$ and integrate to get

$$\int_0^1 y^2\,dx = \lambda \int_0^1 \int_0^1 G(x, \xi)\rho(\xi)y(\xi)y(x)\,d\xi\,dx \tag{7.92}$$

For λ_1, the right-hand side double integral must be as large as possible or have an integrand that does not change sign. Since in the interior $G(x, \xi) > 0$ and $\rho(x) > 0$, it results that $y_1(x)$ must be of the same sign too.

Let x in Eq. (7.91) be the point at which the positive $y_1(x)$ is maximal and $\rho(\xi)$ is a point unit mass at x. Then $1 = \lambda_1 G(x, x)$ and

$$\Phi = 1/\Gamma \tag{7.93}$$

where

$$\Gamma = \max_x \{G(x, x)\} \tag{7.94}$$

In the case of a simply supported circular plate, for instance, $\Gamma = R^2/(16\pi)$ and hence $\Phi = 16\pi/R^2$.

Again, assume that x in Eq. (7.91) is the point at which y_1 is maximal. Application of the mean value theorem, requiring that $G(x, \xi)\rho(\xi) \geqslant 0$, to this equation produces the inequality

$$\lambda_1 \geqslant 1/\max_x \left\{ \int_0^1 G(x, \xi)\rho(\xi)\,d\xi \right\} \tag{7.95}$$

where the integral is recognized to be the static solution to a boundary value problem in which $\rho(x)$ is the load function, and the inequality states that the inverse of the maximum static displacement is less than the fundamental eigenvalue λ_1. The *dynamic* problem of finding the lowest frequency that can be attained with a constant mass variously distributed is transformed thereby into the problem of finding the largest possible *static* displacement caused by a load variously distributed.

6. Maximum Norms and Condition Numbers

By definition, the maximum or l_∞ norm of K is

$$\|K\|_\infty = \max_i \left\{ \sum_{j=1}^{N} |K_{ij}| \right\} \tag{7.96}$$

and it can be computed directly from the entries of K. Otherwise $\|K\|_\infty$ can be estimated from

$$\|K\|_\infty \leqslant \max_e \|k_e\| \eta_{max} \tag{7.97}$$

which should be obvious from the manner in which K is assembled from all k_e.

An upper bound on $\|K^{-1}\|_\infty$ can be established in terms of $\|K^{-1}\|_2$ through the inequality

$$\|K^{-1}\|_\infty \leqslant N^{1/2} \|K^{-1}\|_2 \tag{7.98}$$

where N is the number of rows (columns) in K. The way a bound on $C_\infty(K) = \|K\|_\infty \|K^{-1}\|_\infty$ is formed with Eqs. (7.97) and (7.98) can be seen from the following example: Consider the quadratic string finite element, in which $N = 2Ne + 1$, or approximately $N = 1/h$. The stiffness matrix of this element is given in Eq. (4.42) and we readily have here that

$$\|k_e\|_\infty = \tfrac{16}{3} h^{-1} \tag{7.99}$$

From Eq. (7.45) we have that

$$\|K^{-1}\|_2 = 1/\lambda_1{}^K \leqslant 1/(0.16\pi^2 h) \tag{7.100}$$

and hence

$$C_\infty(K) \leqslant (200/3\pi^2) h^{-2.5} \tag{7.101}$$

which is, at least as far as the exponent -2.5 of h is concerned, too pessimistic.

For problems with a finite influence function, a better upper bound on $\|K^{-1}\|_\infty$ is possible. The derivation of this bound is based on Eq. (5.19), which asserts that the discrete deflection due to a point load, at the point of application, is less than exact. If the unit point force is applied at node number i, then the deflection that it causes at all other nodes is K_{ij}^{-1}, which is $K^{-1}f$ with $f_i = 1$ and $f_{j\neq i} = 0$. At the point of application itself, the response to the unit force is K_{ii}^{-1}, while the exact displacement at this point is $G(x, x)$. Consequently,

$$\max_i \{K_{ii}^{-1}\} \leqslant \Gamma, \qquad \Gamma = \max_x \{G(x, x)\} \tag{7.102}$$

and inasmuch as K is positive definite, $K_{ii}^{-1} > 0$ and therefore

$$\|K^{-1}\|_\infty \leqslant N \max_i \{K_{ii}^{-1}\} \leqslant N\Gamma \tag{7.103}$$

Combination of Eqs. (7.97) and (7.103) yields the desired bound on $C_\infty(K)$:

$$C_\infty(K) \leqslant N\Gamma \max_e \|k_e\|_\infty \eta_{max} \tag{7.104}$$

In the case of a string discretized with quadratic elements, $\Gamma = \frac{1}{4}$, $N = h^{-1}$, and according to Eq. (7.103),

$$\|K^{-1}\|_\infty \leqslant \tfrac{1}{4}h^{-1} \tag{7.105}$$

Then

$$C_\infty(K) \leqslant \tfrac{8}{3}h^{-2} \tag{7.106}$$

which is better than the bound in Eq. (7.101).

The upper bound on $C_\infty(K)$ in Eq. (7.104) fails for the circular membrane; the response y to a unit point force $P(0)$ at $r = 0$, obtained from

$$-\frac{1}{r}\frac{d}{dr}(ry) = P(0), \qquad 0 < r < 1$$
$$y(1) = 0 \tag{7.107}$$

is not bounded at the origin and $\Gamma = \infty$. To observe the dependence of $C_\infty(K)$ upon h in the case of a circular membrane, we resort to a closed form computation of $\|K\|_\infty$ and $\|K^{-1}\|_\infty$. From Eq. (7.107) and

$$E(y, y) = \int_0^1 ry'^2 \, dr \tag{7.108}$$

we get the linear element stiffness matrix

$$k_e = \frac{1}{2}(2e - 1)\begin{bmatrix} 1 & -1 \\ -1 & 1 \end{bmatrix} \tag{7.109}$$

from which

$$K = \frac{1}{2}\begin{bmatrix} 1 & -1 & & & & \\ -1 & 4 & -3 & & & \\ & -3 & 8 & -5 & & \\ & & -5 & 12 & -7 & \\ & & & -7 & 16 & -9 \\ & & & & \cdot & \cdot & \cdot \\ & & & & & \cdot & \cdot & \cdot \\ & & & & & & \cdot \end{bmatrix} \qquad (7.110)$$

is assembled. We readily verify that for this K, $\|K\|_\infty = 4(N-1)$. Some lengthier calculations lead to

$$\max_i \{K_{ii}^{-1}\} = 2\left(1 + \frac{1}{3} + \frac{1}{5} + \cdots + \frac{1}{2N-1}\right) \qquad (7.111)$$

and when N is large

$$\max_i \{K_{ii}^{-1}\} = \log N \qquad (7.112)$$

In the same way, we find also that

$$\|K^{-1}\|_\infty = 2\left(1 + \frac{2}{3} + \frac{3}{5} + \cdots + \frac{N}{2N-1}\right) \qquad (7.113)$$

and then

$$C_\infty(K) = 4N^2 + 2N \log N \qquad (7.114)$$

Even though, according to Eq. (7.112), the diagonal of K^{-1} grows like $\log N$, $C_\infty(K)$ is still $O(N^2)$.

7. Scaling

Scaling is the computationally inexpensive congruent transformation $K' = DKD$ with a diagonal D to improve some properties of K' over that of K. Often the purpose of scaling is to improve the condition of the global stiffness matrix, which is said to be *optimally scaled* with respect to the spectral condition number when

$$C_2(K') = \min_D \{C_2(DKD)\} \qquad (7.115)$$

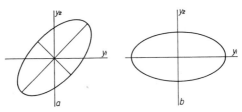

Fig. 7.5 Energy ellipsoid in two dimensions with major axes (eigenvectors) (a) inclined 45° to the axes, and (b) parallel to the axes.

A general procedure for creating an *optimal* D, without excessive computation, is not known, but the choice $D_{ii} = 1/K_{ii}^{1/2}$, which makes all diagonal entries of K equal to 1, often does well in positive definite matrices. Why and when is disclosed in Fig. 7.5, which shows the contours of the energy quadratic $y^T K y$ for a 2×2 matrix. Scaling such that $(DKD)_{ii} = 1$ is geometrically construed as making the energy ellipsoid cut unit segments on the $y_1, y_2, \ldots,$ y_N axes. The ellipse in Fig. 7.5a, with eigenvectors inclined 45° to the axes, already cuts nearly equal segments on them, and scaling hardly changes the intrinsic form of the ellipse. Consequently $C_2(DKD)$ is only slightly affected by it. For the ellipse in Fig. 7.5b, the situation is entirely different. Its eigenvectors are parallel to the axes, scaling reduces the ellipse to a circle, and the condition of K is maximally improved thereby.

Thus, we see that for scaling to be effective, the eigenvectors, particularly those corresponding to the extremal eigenvalues, ought to be of the form $(0, 0, \ldots, 1, \ldots, 0, 0)$ or close to it. In case the eigenvectors are not close to the axes, a preliminary rotation of the axes to be followed by scalling is helpful. A practical example of this is provided by the elastic tube shown in Fig. 7.6. It consists of an inner annular portion with a low elastic modulus and a rigid envelope of a much higher elastic modulus, such as may be the case in a rocket motor having a soft solid propellant and a tough metal outer casing.

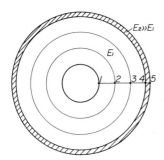

Fig. 7.6 Elastic tube with soft core of elastic modulus E_1 and stiff outer shell of modulus $E_2 \gg E_1$ discretized with four linear elements.

Specifically, let the tube in Fig. 7.6 be discretized with first-order elements, with the typical one situated in the range $r_1 \leqslant r \leqslant r_2$, and of elastic modulus E. Then

$$k_e = \frac{1}{2}E(r_1 + r_2)\begin{bmatrix} 1 & -1 \\ -1 & 1 \end{bmatrix} + \frac{2Eh}{r_1 + r_2}\begin{bmatrix} 1 & \\ & 1 \end{bmatrix} \tag{7.116}$$

in which E for the outer shell is much larger than for the core. It is clear from the physics of this problem that a large amount of elastic energy is stored in the composite tube by the relative displacement of y_4 against y_5 and hence the eigenvector corresponding to this large eigenvalue is nearly $(0, 0, 0, 1, -1)$. An energy ellipsoid as in Fig. 7.5a corresponds to this, and scaling is of little avail. Rotation corrects this situation and, indeed with the transformation

$$z_4 = \tfrac{1}{2}(y_5 + y_4), \qquad z_5 = \tfrac{1}{2}(y_5 - y_4) \tag{7.117}$$

the eigenvector corresponding to the largest energy expenditure becomes $(0, 0, 0, 0, 1)$ and scaling is successful in improving the condition of K.

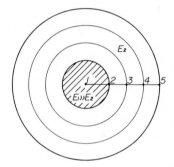

Fig. 7.7 Elastic composite disk with soft outer ring of elastic modulus E_2 and stiff inner core of elastic modulus $E_1 \gg E_2$ discretized with four linear elements.

To gain further insight into the scaling mechanism of K, it is instructive that we also consider the disk in Fig. 7.7, which now has an *inner stiff* core and an *outer soft* shell. We have no difficulty in finding out that the largest energy expenditure in this case comes from $y_2 - y_1$, but here $y_1 = 0$ and hence the eigenvector corresponding to the largest eigenvalue is actually close to $(1, 0, 0, 0)$. In this case, scaling is effective without previous rotation.

We encountered scaling earlier when we changed the nodal value y' into hy' in order to remove h from the interior of the element stiffness and mass matrices in Eqs. (4.47) and (4.53). Let us first consider the beam element. Without scaling and with h left inside k_e and m_e in Eqs. (4.47) and (4.53), we have that $\lambda_1^{m_e} = O(h^3)$, $\lambda_4^{k_e} = O(h^{-3})$, and consequently Eq. (7.40) makes the prediction that $C_2(K) \leqslant O(h^{-6})$. With scaling that removes h from the interior of both k_e and m_e, we have that $\lambda_1^{m_e} = O(h)$, $\lambda_4^{k_e} = O(h^{-3})$, and

Fig. 7.8 Maximal response Γ to a unit force and unit torque acting on a cantilever beam.

Eq. (7.40) makes the different prediction that $C_2(K) \leqslant O(h^{-4})$. Numerical computations insist that the beam discretized with the element stiffness matrix in Eq. (4.53) be $C_2(K) = O(h^{-4})$, both with and without scaling, and the prediction of Eq. (7.40) is too pessimistic when h is left inside the element matrices. Our deceptive bound $O(h^{-6})$ unfortunatly supports a prevailing popular wisdom, maintaining that a matrix with such wide ranging diagonal terms as $O(1)$ and $O(h^2)$ is, per force, more ill conditioned than one with only $O(1)$ terms on its diagonal. The fallacy of this argument for the cubic beam element can be proved with the aid of the bound on $C_\infty(K)$ in Eq. (7.104). It is obvious that for k_e in Eq. (4.53)

$$N\|k_e\|_\infty = O(h^{-4}) \qquad\qquad (7.118)$$

whether k_e is scaled or not. What about Γ? Since the nodal values also include, in addition to the translation y, the rotation y', Γ in Eq. (7.104) stands for the maximal displacement or rotation response to either a point unit force or a point unit *torque*, respectively, as in Fig. 7.8. It happens that the beam has a finite response to both, $\Gamma < \text{const.}$ for any h, and $C_\infty(K) \leqslant O(h^{-4})$ even without scaling. Because $C_2(K) \leqslant C_\infty(K)$, the same is also true for the spectral condition number $C_2(K)$, which is also $O(h^{-4})$ with or without scaling.

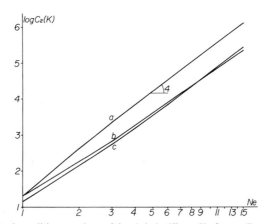

Fig. 7.9 Spectral condition numbers of the global stiffness K of a cantilever beam assembled from cubic C^1 elements. (a) Unscaled K, (b) K scaled by putting $h = 1$ inside the element matrices, and (c) K scaled to $K_{ii} = 1$.

Figure 7.9 shows the variation of $C_2(K)$ with the number of elements for a cantilever beam [i.e., $y'''' = P(1)$ $0 < x < 1$, $y(0) = y'(0) = y''(1) = 0$] discretized with cubic elements with a scaled and unscaled global stiffness matrix K, and indeed in both cases $C_2(K)$ is proportional to Ne^4. In fact, for the unscaled K, $C_2(K) = 10.5Ne^4$, while when $K_{ii} = 1$, $C_2(K) = 5.3Ne^4$. The $C_\infty(K)$ of this beam is numerically found to be about 10 times larger than $C_2(K)$.

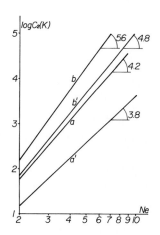

Fig. 7.10 Condition of the global stiffness matrix K of a circular plate discretized with cubic finite elements. Curves a and a' are for a uniform mesh without and with scaling, respectively, while b and b' are for a mesh graded according to $h_e = e^{1.5}h_{min}$.

The analysis of the previous section predicts that the circular plate, discretized with a nonuniform mesh of finite elements, has a global stiffness matrix K for which $C_2(K) \leqslant C_\infty(K) \leqslant O(Neh_{min}^{-2})$, where h_{min} is the smallest element near the center of the plate. Figure 7.10 shows the actual variation of $C_2(K)$ for the circular plate that is discretized once with a uniform mesh and then with a mesh graded by $h_e = e^{1.5}h_{min}$. Curves a and a' refer to the uniform mesh without and with scaling, respectively, while curves b and b' refer to the graded mesh with and without scaling, respectively. For a given number of elements, the graded mesh produces a more ill conditioned matrix than that obtained from the uniform mesh, but the purpose of mesh grading is to save on the number of elements while aiming at a given discretization accuracy.

We can prove that with cubic beam elements that include as nodal values y and y', $C_2(K)$ is proportional to Ne^4 even without scaling because the beam can carry both a point force and a point torque. A string discretization with elements including y' among the nodal values is different. If we use Eq. (7.40) to bound $C_2(K)$ while leaving h inside k_e and m_e in Eq. (4.47), we get too pessimistic results as happened for the beam. Circumventing this difficulty with Eq. (7.104) to bound $C_\infty(K)$ as we did for the beam is no longer

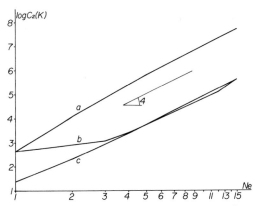

Fig. 7.11   Same as Fig. 7.9 but this time assembly of K is from C^2 quintic elements.

possible here because the Γ corresponding to y' is *infinite* for the string—the string cannot carry a torque. To satisfy our curiosity as to the effect of scaling on $C_2(K)$ when K is assembled from higher-order elements with higher-order nodal values, we revert to numerical computations and consider in particular the cantilever beam discretized with C^2 quintic elements for which the element matrices are

$$
k_e = \frac{1}{70h^3}
\begin{bmatrix}
1200 & 600h & 30h^2 & -1200 & 600h & -30h^2 \\
600h & 384h^2 & 22h^3 & -600h & 216h^2 & -8h^3 \\
30h^2 & 22h^3 & 6h^4 & -30h^2 & 8h^3 & h^4 \\
-1200 & -600h & -30h^2 & 1200 & -600h & 30h^2 \\
600h & 216h^2 & 8h^3 & -600h & 384h^2 & -22h^3 \\
-30h^2 & -8h^3 & h^4 & 30h^2 & -22h^3 & 6h^4
\end{bmatrix}
$$

$$
m_e = \frac{h}{55440}
\begin{bmatrix}
21720 & 3732h & 281h^2 & 6000 & -1812h & 181h^2 \\
3732h & 832h^2 & 69h^3 & 1812h & -532h^2 & 52h^3 \\
281h^2 & 69h^3 & 6h^4 & 181h^2 & -52h^3 & 5h^4 \\
6000 & 1812h & 181h^2 & 21720 & -3732h & 281h^2 \\
-1812h & -532h^2 & -52h^3 & -3732h & 832h^2 & -69h^3 \\
181h^2 & 52h^3 & 5h^4 & 281h^2 & -69h^3 & 6h^4
\end{bmatrix}
$$

$$\text{(7.119)}$$

relative to the nodal values $y_1, y_1', y_1'', y_2, y_2', y_2''$. The variation of $C_2(K)$ with the number of elements Ne for this problem is traced in Fig. 7.11 for unscaled K and for two kinds of scaling: one achieved by setting $h = 1$ inside the element matrices in Eq. (7.119), which is equivalent to the choice of nodal values $y_1, hy_1', h^2y_1'', y_2, hy_2', h^2y_2''$, and then with one that makes $K_{ii} = 1$. It is interesting that even with this higher-order element, $C_2(K)$ is still proportional to Ne^4 even without scaling, but a much larger reduction in the numerical values of the condition number has been reached here than for the cubic element. For the quintic beam element, $C_2(K) = 1060Ne^4$ without scaling, and $C_2(K) = 8.4Ne^4$ with scaling.

Up until now, we have only considered the effect of scaling on $C_2(K)$—the ratio between the extremal eigenvalues of the global stiffness matrix K. But even if the scaling that makes all $K_{ii} = 1$ does not much change the condition of K, it can still profoundly and beneficially affect the complete distribution of the eigenvalues of the global stiffness matrix. To see how, consider the beam discretized with the cubic element in Eq. (4.53). We have just argued that the assembled global stiffness matrix is $O(h^{-4})$ whether K is scaled or not. What happens to the complete spectrum of K with scaling is shown in Fig. 7.12. This figure plots the normalized eigenvalues $\lambda_j/\lambda_1, j = 1, 2, \ldots, N$, of K for a simply supported beam, with each graph consisting of two branches corresponding to the two different shape functions for the displacements and the rotations. Only the symmetric modes of this beam are considered and the curves in Fig. 7.12 are made for the beam with 17 elements for its half, once unscaled and once scaled to $K_{ii} = 1$. In the first case, $C_2(K) = 10^{6.8}$ while with scaling the spectral condition number is reduced to $10^{5.7}$. The effect of scaling on the eigenvalues distribution is more dramatic—it nearly flattens the higher branch *rendering half of the eigenvalues of K almost equal*. Certain iterative procedures used to solve the algebraic equilibrium equation $Ky = f$ benefit greatly from this. The method of conjugate gradients, for

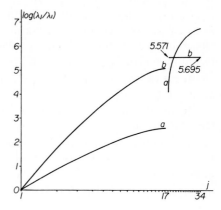

Fig. 7.12 Variation of λ_j/λ_1 in a (half) simply supported beam [i.e., $y(0) = y''(0) = y'(\tfrac{1}{2}) = y'''(\tfrac{1}{2}) = 0$] discretized with 17 finite elements when (a) K is unscaled and (b) K is scaled.

instance, converges faster in the presence of nearly equal eigenvalues in K, and here scaling has the effect of cutting the solution effort by nearly one half.

8. Positive Flexibility Matrices

A matrix whose entries are all positive is said to be *positive*. Some global stiffness matrices generated by finite elements for second-order problems possess positive inverses, or *flexibility* matrices. We recognize them and learn some of their basis properties by the following theorems:

(i) *If the positive definite stiffness matrix K is such that $K_{ij} \leqslant 0$ for all $i \neq j$ (i.e., K is a Stieltjes matrix), then the flexibility matrix K^{-1} is positive:* $K_{ij}^{-1} > 0.$

(ii) *The eigenvector y_1 corresponding to $\lambda_1{}^K$ (or $\lambda_N^{K^{-1}}$) is positive when $K^{-1} > 0.$*

(iii) *When the positive definite and symmetric stiffness matrix K is tridiagonal with negative off diagonal entries, the eigenvector corresponding to $\lambda_N{}^K$ or $\lambda_1^{K^{-1}}$ has alternating signs.*

From the way K is assembled from the element matrices, it is obvious that $(k_e)_{ij} \leqslant 0, i \neq j$, implies that $K_{ij} \leqslant 0$ when $i \neq j$. Hence, to decide the positivity of K^{-1}, it is enough to examine only the typical finite element stiffness matrix. We see, for instance, that the first-order lumped string element for which

$$k_e = \frac{1}{h} p_c \begin{bmatrix} 1 & -1 \\ -1 & 1 \end{bmatrix} + \frac{1}{2} h q_c \begin{bmatrix} 1 & \\ & 1 \end{bmatrix} \tag{7.120}$$

with $p_c > 0$ and $q_c \geqslant 0$, gives rise to $K^{-1} > 0$.

The implication of a positive flexibility matrix K^{-1} is that a positive point force causes only positive displacements and that the effect of this load is felt *at all nodal points*. The global stiffness matrix K is *sparse* but the global flexibility matrix K^{-1} is *dense* and this has important repercussions on the storage and inversion methods used to solve the linear algebraic stiffness equation $Ky = f$.

The computational interest in positive flexibility matrices stems from the fact that when $K^{-1} > 0$, *numerical* bounds of arbitrary closeness can be computed for $C_2(K)$ and $C_\infty(K)$. To construct such bounds on $\lambda_1{}^K$, let $y > 0$ be a sequence of vectors converging to the here positive fundamental eigenvector y_1. Let y_i, $i = 1, 2, \ldots, N$, be the entries of the vector y and choose $D_{ii} = y_i$ in the *similarity transformation* $D^{-1}KD$ that leaves all eigenvalues of K and its sign pattern unchanged. Then

$$(D^{-1}KDe)_i = (Ky)_i/y_i, \qquad e^T = (1, 1, \ldots, 1) \tag{7.121}$$

and consequently, because $K_{ij} \leq 0$, $i \neq j$, Gerschgorin's and Rayleigh's theorems enclose λ_1^K in

$$y^T K y/y^T y \geq \lambda_1^K \geq \min_i \{(Ky)_i/y_i\} \tag{7.122}$$

and as $y \to y_1$, the lower and upper bounds in Eq. (7.122) close upon λ_1^K.

Concerning $\|K^{-1}\|_\infty$, since $K_{ij}^{-1} > 0$, we have that $\|K^{-1}\|_\infty = \|K^{-1}e\|_\infty$ and therefore if $r = Ky - e$, then

$$\frac{\|y\|_\infty}{1 + \|r\|_\infty} \leq \|K^{-1}\|_\infty \leq \frac{\|y\|_\infty}{1 - \|r\|_\infty} \tag{7.123}$$

provided that $\|r\|_\infty < 1$.

As an example of the computational bounds on the norms and condition of K when $K^{-1} > 0$, consider the global stiffness matrix K in Eq. (7.110) for the circular membrane. We let K be of size 25×25 and wish to bound λ_1^K and λ_{25}^K of it, using Eq. (7.122) and

$$y^T K y/y^T y \leq \lambda_N^K \leq \max_i \{(Ky)_i/y_i\} \tag{7.124}$$

in which y has alternating signs. To generate the converging sequence of y vectors in Eq. (7.122), we use inverse iteration starting with $y^T = (25, 24, \ldots, 1)$. Because $K^{-1} > 0$, all improved y vectors remain positive, as required in Eq. (7.122), and converge to y_1. To generate the converging sequence of y vectors for Eq. (7.124), we use the power method, starting with $y^T = (1, -1, 1, -1, \ldots)$. Because of the special sign distribution in K, all subsequent vectors remain with alternating signs and converge to y_N. The results of this computation are summarized in the accompanying table for K in Eq. (7.110) but without the $\frac{1}{2}$ in front of it.

	λ_1		λ_{25}		$C_2(K)$	
Step	Upper	Lower	Upper	Lower	Upper	Lower
1	0.1131	0.0400	184.0	94.12	4600	832
2	0.1072	0.0936	176.1	140.7	1881	1313
3	0.1069	0.1040	173.0	153.8	1663	1439
4	0.1069	0.1063	171.0	159.5	1609	1492
5	0.1069	0.1068	169.8	162.4	1590	1519
	exact		exact		exact	
	0.1069		166.8		1560	

9. Computational Errors

Computational errors issue from two main sources: the storage of the original global stiffness matrix and the load vector in a finite word length computer, and the solution procedure for $Ky = f$ that operates arithmetically with round-off. Experience suggests that direct solution methods like Gauss elimination create computational errors in the solution comparable in magnitude to those occurring initially when the data are stored, and that the total computational error δy in y in a computer with s significant digits is adequately given by

$$\|\delta y\|_2 / \|y\|_2 = C_2(K)10^{-s} \tag{7.125}$$

where $C_2(K)$ is the spectral condition number of the *optimally scaled K*. On a computer with s significant digits, numerical singularity occurs in K when $C_2(K) = 10^s$.

Our previous discussion of the condition of the finite element global stiffness matrix assures us that modern computers that typically carry more than six significant digits are capable of solving any reasonable finite element problem with tolerable round-off errors. Fourth-order problems give rise to stiffness matrices that become ill conditioned at a faster rate than those arising in second-order problems. In this case we might be required to use double precision, which is only some 10% more time consuming than single precision, but which requires twice the computer storage, or we might use higher-order elements of which fewer are required for any given discretization accuracy.

The cantilever beam discretized with cubic and quintic finite elements, whose condition numbers are shown in Figs. 7.9 and 7.11, serves well to illustrate the reduction in round-off errors for a fixed discretization accuracy through the use of higher-order elements. We recall that with cubic elements $C_2(K) = 10.54Ne^4$ and $C_2(K) = 5.31Ne^4$ when K is unscaled and scaled, respectively, while with quintic elements the corresponding condition numbers and $C_2(K) = 1066Ne^4$ and $C_2(K) = 8.4Ne^4$. Since the *scaled* spectral condition numbers are nearly equal for a given number of elements for both discretizations, we also expect the computational errors to be nearly equal for the same number of elements in the two different discretizations. Figures 7.14 and 7.15 show the error, which is here purely computational (the finite element approximation being capable of duplicating exactly the cubic solution) in the tip deflection of a point-loaded cantilever beam discretized with cubic and quintic finite elements and solved by Gauss elimination. Computations were undertaken on a computer with about seven significant digits and we see that in both cases the total computational error grows as $12.6Ne^4 \times 10^{-7}$ or about $C_2(K) \times 10^{-7}$ where $C_2(K)$ is the scaled spectral

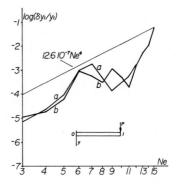

Fig. 7.13 Computation error in the tip deflection of a cantilever beam discretized with cubic C^1 finite elements when (a) K is unscaled and (b) K is scaled. Solution of $Ky = f$ is done using Gauss elimination on a computer with seven significant digits.

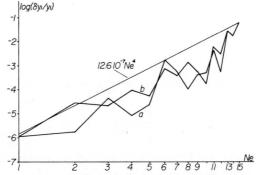

Fig. 7.14 Same as Fig. 7.13 but K is created from quintic C^2 elements.

condition number of both Ks. Hence, for the same number of elements, the round-off or computational errors are the same whether K is assembled from cubic or quintic elements, but a greater accuracy is possible with the quintic element of which fewer are needed to arrive at a certain prescribed discretization accuracy. Figures 7.13 and 7.14 show the errors in the y nodal points only; the computational errors in the nodal y' and y'' are the same.

To observe the computational errors that occur in eigenproblems, we undertake to calculate the fundamental frequency of an elastic disk fixed around its rim, given by

$$r^2 y'' + r y' + (r^2 \lambda - 1) y = 0, \qquad y(1) = 0 \qquad (7.126)$$

Here the first eigenfunction is $y = J_1(\sqrt{\lambda} r)$ and λ_1 is obtained from the boundary condition $J_1(\sqrt{\lambda_1}) = 0$ or $\lambda_1 = 14.682$. Using linear finite elements, we set up the algebraic eigenproblem $Ky = \lambda M y$, for which we know $C_2(K) = O(Ne^2)$ and $C_2(M) = O(1)$. High and low precision computations are used to get the computational error $\delta \lambda_1$ in λ_1 shown in Fig. 7.15. A Jacobi method is employed for the extraction of the eigenvalues and the last computation is done with seven significant digits. Again, the computational errors are

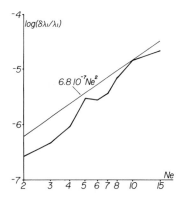

Fig. 7.15   Computational error $\delta\lambda_1/\lambda_1$ in the fundamental eigenvalue of an elastic disk fixed at its rim. Jocobi's method is used with seven significant digits.

proportional to the machine accuracy and the condition of K, which is here $C_2(K) = 4.4Ne^2$; $C_2(M)$ is a constant independent of h.

10.   Detection of Computational Errors

We accept from the discussion of the preceding sections that it is prohibitively expensive to obtain a concrete numerical bound on $C_2(K)$ for any particular global stiffness matrix K, or on the computational errors that occur with any particular solution method. More sophisticated routines to invert K or solve $Ky = f$ include a check on the pivot in each elimination stage and they exit in case a pivot is detected at the level of the machine accuracy, signaling numerical singularity or if the pivot changed its sign. But if the stiffness matrix is not ill conditioned to the point of numerical singularity, the solution procedure naturally terminates, or stops if iterative, leaving us with the dilemma of estimating the computational errors in the solution.

One way to assess these errors is to entirely solve the algebraic system *twice*, once in single and then in double precision, and to observe the changing digits. Another way is to invent a solution y, compute f for it from $f = Ky$, then recover y from the solution of $Ky = f$, and compare it with the original choice in order to estimate $C_2(K)$. Or *iterative improvement* may be used to this end.

The underlying idea of iterative improvement is to solve the algebraic system in single precision and improve the solution, using double (higher) precision calculations. In order to describe iterative improvement, let the original global stiffness matrix be K and the perturbed one, the one that produces the erroneous solution, be $K + E$. Also, let $F = (K + E)^{-1}$. Forgetting for the moment round-off errors, or assuming sufficiently high precision computations, the iterative procedure

$$y^{(1)} = (I - FK)y^{(0)} + Ff \qquad (7.127)$$

generates a sequence of vectors $y^{(0)}$, $y^{(1)}$, $y^{(2)}$, ... that converges to $y = K^{-1}f$ if $\|I - FK\| < 1$. Often F is not explicitly created, but $y^{(0)}$ is obtained from the solution of $(K + E)y^{(0)} = f$ by, say, the triangular factorization of $K + E$. Introduction of the residual $r^{(0)} = Ky^{(0)} - f$ is thus appropriate, and Eq. (7.127) is written as

$$y^{(1)} = y^{(0)} - Fr^{(0)} \qquad (7.128)$$

In the iterative improvement, $r^{(0)} = Ky^{(0)} - f$ is computed in the *highest precision* from K and f stored in the same precision. Next, $(K + E)(F^{(0)}r^{(0)}) = r^{(0)}$ is solved in the lower precision used to get $y^{(0)}$, and the improved $y^{(1)}$ is formed according to Eq. (7.128). The success of iterative improvement depends on the E caused by round-off error being small enough for $\|I - FK\| < 1$ to hold. Then the change in $y^{(1)}$ over $y^{(0)}$ indicates the amount of round-off errors in it and if convergence is fast, $10^s\|y^{(1)} - y^{(0)}\|_2/\|y^{(1)}\|_2$, where s is the number of significant digits in the single precision computations, provides a good approximation to $C_2(K)$. The seeming advantage of this more complicated procedure is that $K + E$ need be decomposed or inverted only once in lower precision, while the computation of $r = Ky - f$ can be done in high precision directly on the element data, which may often easily be stored in the finite element method in higher precision.

EXERCISES

1. Discretize the cantilever beam in Fig. 7.16 with two cubic finite elements and compute $C_2(K)$ as a function of $p_1 > 1$ without and with scaling that renders all $K_{ii} = 1$. Explain your results.

Fig. 7.16 Two-section cantilever beam with stiff root section and soft tip section.

2. Discretize the uniform unit cantilever beam with quintic finite elements as given in Eq. (7.119). Assume its deflection to be of the form $y = x^2$, $0 \leqslant x \leqslant 1$, and compute the corresponding load vector f from $f = Ky$. Solve $Ky = f$ with this f and recover y. Use single precision throughout and repeat your computations with a scaled K. Compare the assumed y to the computed and estimate $C_2(K)$ from Eq. (7.125). See how close you get to $C_2(K) = 8.4Ne^4$.

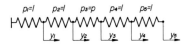

Fig. 7.17 Five springs in series with a fixed left-hand end point and a free right-hand end point. All spring constants equal 1 except for the third which has stiffness p.

3. Consider the spring series in Fig. 7.17, which is attached on the left and free on the right. The element stiffness matrix of the typical eth spring of constant p_e is

$$k_e = p_e \begin{bmatrix} 1 & -1 \\ -1 & 1 \end{bmatrix} \qquad (7.129)$$

Use this k_e to assemble the global stiffness matrix for the spring system and study its condition when $p \to 0$ and when $p \to \infty$. Discuss the transformations that cure the ill conditioned K when $p \gg 1$ or $p \ll 1$.

4. Use the cubic string element stiffness matrix in Eq. (4.47) to assemble K for the fixed unit string and study the effect of scaling on this $C_2(K)$. The element stiffness and mass matrices for the quintic string element with nodal values $y_1, y_1', y_1'', y_2, y_2', y_2''$, are given by

$$k_e = \frac{1}{1260h} \begin{bmatrix} 1800 & 270h & 15h^2 & -1800 & 270h^2 & -15h^2 \\ 270h & 288h^2 & 21h^3 & -270h & -18h^2 & 6h^3 \\ 15h^2 & 21h^3 & 2h^4 & -15h^2 & -6h^3 & h^4 \\ -1800 & -270h & -15h^2 & 1800 & -270h & 15h^2 \\ 270h & -18h^2 & -6h^3 & -270h & 288h^2 & -21h^3 \\ -15h^2 & 6h^3 & h^4 & 15h^2 & -21h^3 & 2h^4 \end{bmatrix}$$

$$m_e = \frac{h}{55440} \begin{bmatrix} 21720 & 3732h & 281h^2 & 6000 & -1812h & 181h^2 \\ 3732h & 832h^2 & 69h^3 & 1812h & -532h^2 & 52h^3 \\ 281h^2 & 69h^3 & 6h^4 & 181h^2 & -52h^3 & 5h^4 \\ 6000 & 1812h & 181h^2 & 21720 & -3732h & 281h^2 \\ -1812h & -532h^2 & -52h^3 & -3732h & 832h^2 & -69h^3 \\ 181h^2 & 52h^3 & 5h^4 & 281h^2 & -69h^3 & 6h^4 \end{bmatrix}$$

Use this k_e to repeat the above study.

5. Scale the global stiffness matrix

$$K = \begin{bmatrix} 2 & -1 & \\ -1 & 2 & -1 \\ & -1 & 2 \end{bmatrix}$$

with the diagonal matrix $D_{11} = 1$, $D_{22} = \alpha$, $D_{33} = 1$. Show that

$$C_2(DKD) = \frac{1 + \alpha^2 + (1 + \alpha^4)^{1/2}}{1 + \alpha^2 - (1 + \alpha^4)^{1/2}}$$

and that its minimum occurs at $\alpha = 1$.

6. Consider the linear system $Ky = f$ with

$$K = \begin{bmatrix} 2 & -1 \\ -1 & 2 \end{bmatrix}, \qquad f = \begin{bmatrix} 1 \\ 4 \end{bmatrix}$$

with the solution $y^T = (2, 3)$. Use the transformation $y = Dx$ with the diagonal $D_{11} = 1$, $D_{22} = |\alpha|$ to get $K'x = f'$, $K' = DKD$, $f' = Df$. Discuss the effect of $|\alpha|$ on the accuracy of the solution of $K'x = f'$ in the presence of round-off errors when $|\alpha| \ll 1$.

7. Use cubic finite elements to discretize the eigenproblem $y'''' = \lambda y''$, $0 < x < 1$, $y(0) = y''(0) = y(1) = y''(1)$, describing the stability of a column. The corresponding algebraic eigenproblem is $Ky = \lambda K'y$ in which $C_2(K) = O(h^{-4})$ and $C_2(K') = O(h^{-2})$. Establish numerically the dependence of the computational error $\delta\lambda_1/\lambda_1$ on the number of elements and in turn on the condition of K and K'.

8. Use the Ritz method to solve $-y'' = 2$, $0 < x < 1$, $y(0) = y'(1) = 0$ with $\phi_j = x^{1+(j-1)\varepsilon}$ $j = 1, 2, \ldots, N$ and $\varepsilon = 1/(N-1)$ to assure the presence of x and x^2 in $\tilde{y}$, and hence only *computational* errors. Show that here *both* K and M become fast ill conditioned as N increases. When $N = 6$ they are practically indefinite with seven significant figures. Compute the weights of the trial function $\tilde{y} = a_1\phi_1 + a_2\phi_2 + \cdots$ to show that they come out very inaccurate but that $\hat{y}$ is very good even when K is numerically indefinite. Contrast this situation with finite elements.

SUGGESTED FURTHER READING

Varga, R. S., *Matrix Iterative Analysis*. Prentice Hall, Englewood Cliffs, New Jersey, 1962.
Fadeev, D. K., and Fadeeva, V. N., *Computational Methods of Linear Algebra*. Freeman, San Francisco, California, 1963.
Roach, G. F., *Green's Function; Introductory Theory with Applications*. Van Nostrand Reinhold, Princeton, New Jersey, 1970.

8 *Equation of Heat Transfer*

1. Nonstationary Heat Transfer in a Rod

The equation whose numerical solution we shall consider in this chapter describes many diffusion, conduction, and decay phenomena. We prefer to tie our discussion specifically to heat conduction because of its prevalence and also because of the tangible physical nature of the coefficients and variables involved in this process. To acquire intuition into the physical nature of what we analyze mathematically, we shall also derive in detail the equation of nonstationary heat conduction in a rod from the basic physical laws. Afterward, we shall proceed to demonstrate and analyze the various numerical techniques for its solution.

Consider, then, a thin heat conducting rod of variable cross section a, as shown in Fig. 8.1, unevenly heated to the temperature y, which, because of the rod's thinness, we assume to be uniform throughout the cross section, and loosing heat by radiation to an environment of zero temperature. Because of the original variable temperature distribution in the rod, which does not generate heat by itself, heat flows from the hot regions of the rod into the cold, causing the temperature distribution to change not only along x but also with the time t. The equation formulating the variation of the

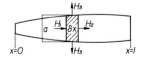

Fig. 8.1 Thin heat conducting rod of variable cross section a, with control section δx.

158

temperature $y(x, t)$ with space and time is derived from the first law of thermodynamics, which stresses that heat is a form of energy, conserved throughout the conduction process. To establish the local heat balance, a slice of the rod is isolated, as in Fig. 8.1, and the heat entering it, leaving it, and remaining inside it is counted. As is customary, we denote by H the *rate* of heat entering the control volume and have that at station 1, H_1 enters it by conduction, at station 2, H_2 flows out of it in the same way, while on the surface of the rod, H_3 escapes from it by radiation. The net result of this flow is a heat buildup (loss) inside the control volume that induces a local temperature rise (drop) at the rate of

$$\rho c a \, \delta x \frac{\partial y}{\partial t} = H_1 - H_2 - H_3 \tag{8.1}$$

in which ρ denotes the density of the rod and c its *specific heat*. Fourier's law relates the heat flow H to the temperature gradient, such that at cross sections 1 and 2 we have from this that

$$H_1 = -\left(k a \frac{\partial y}{\partial x}\right)_1 \qquad H_2 = -\left(k a \frac{\partial y}{\partial x}\right)_2 \tag{8.2}$$

where k is the *thermal conductivity* of the rod. The heat losses through radiation are, according to Newton's law of cooling, proportional to the temperature difference between the rod's skin and the ambient temperature, or here

$$H_3 = q y \, \delta x \tag{8.3}$$

where $q(x)$ is the heat transfer coefficient per unit length of the rod. Substituting Eqs. (8.2) and (8.3) in the heat balance equation (8.1) and proceeding to the limit $\delta x \to 0$ produce the heat equilibrium equation

$$-\frac{\partial}{\partial x}\left(k a \frac{\partial y}{\partial x}\right) + q y = -\rho c a \frac{\partial y}{\partial t}, \qquad 0 < x < 1, \qquad t > 0 \tag{8.4}$$

that we sought to establish.

All the thermal and geometrical parameters $a > 0$, $k > 0$, $q \geqslant 0$, $\rho > 0$, and $c > 0$ in Eq. (8.4) may be functions of x. They may also be only piecewise continuous as when the cross section a of the rod experiences a discontinuity or when the rod is made up of segments of different thermal properties. But it follows from thermal equilibrium that across any section of the rod, both the temperature y and the heat flux $k a y'$ are always continuous.

At $t = 0$, the temperature $y(x, t)$ has its initial distribution $y(x, 0)$—the *initial condition* for Eq. (8.4). At the end points of the rod, the *boundary conditions* are generally of the form

$$kay' = h(y - y_a) + H, \qquad \text{at} \quad x = 0, \quad t > 0$$
$$-kay' = h(y - y_b) - H, \qquad \text{at} \quad x = 1, \quad t > 0$$

(8.5)

in which h is the heat transfer coefficient at the ends, and y_a and y_b the ambient temperatures at $x = 0$ and $x = 1$, respectively. When $h = \infty$, the boundary conditions in Eq. (8.5) become that of the prescribed temperature, when $h = 0$, they become that of the prescribed heat flux H into the rod, and when $H = 0$, they become the radiation boundary conditions.

Equation (8.4), together with the initial temperature distribution $y(0, x)$ and the boundary conditions in Eq. (8.5), constitutes a *linear first-order initial value problem.*

If the rod is assumed to be insulated along its entire length, such that $q(x) = 0$, of constant cross section and thermal coefficients, Eq. (8.4) simplifies to

$$y'' = \alpha \dot{y}$$

(8.6)

where α is the *thermal diffusivity* of the rod and $\dot{y} = \partial y / \partial t$.

Most of what is important and special in the numerical solution of the first-order heat conduction initial value problem can be conveniently argued and displayed in the simpler equation (8.6).

2.   Finite Difference Approximation

Finite elements or finite differences discretize the boundary value problem by reducing the differential equation and boundary conditions to a system of (linear) algebraic equations. Discretization *in space only* of the partial differential equation (8.4) reduces it to a *system of ordinary differential equations in time.* In this reduction, Eq. (8.4) or (8.6) is considered instantaneously at time t, and if we choose to replace y'' by a central scheme, then for the typical interior jth node, the approximate finite difference equation for Eq. (8.6) is

$$h^{-2}(-y_{j-1} + 2y_j - y_{j+1}) + \alpha \dot{y}_j = 0$$

(8.7)

in which all the y nodal values are functions of time. If we further assume the rod to be held at a constant zero temperature at its ends, such that $y(0, t) = y(1, t) = 0$, $t > 0$, then also $\dot{y}(0, t) = \dot{y}(1, t) = 0$. Consequently, in Eq. (8.7), $y_0 = y_{N+1} = \dot{y}_0 = \dot{y}_{N+1} = 0$ and the coupled system of first-order equations

$$\frac{1}{h^2}\begin{bmatrix} 2 & -1 & & & & \\ -1 & 2 & -1 & & & \\ & \cdot & \cdot & \cdot & & \\ & & \cdot & \cdot & \cdot & \\ & & & \cdot & \cdot & \cdot \\ & & -1 & 2 & -1 \\ & & & -1 & 2 \end{bmatrix}\begin{bmatrix} y_1 \\ y_2 \\ \cdot \\ \cdot \\ \cdot \\ y_{N-1} \\ y_N \end{bmatrix} + \begin{bmatrix} \dot{y}_1 \\ \dot{y}_2 \\ \cdot \\ \cdot \\ \cdot \\ \dot{y}_{N-1} \\ \dot{y}_N \end{bmatrix} = 0 \qquad (8.8)$$

results, which we write concisely as

$$Ky + \dot{y} = 0 \qquad (8.9)$$

where the vector y is a function of time.

Finite difference discretization of the more general heat conduction problem in Eqs. (8.4) and (8.5) leads to the same system of equations as (8.9), whose numerical integration in time is the subject of this chapter.

3. Modal Analysis

This is the discrete counterpart to eigenfunction expansion or separation of variables and has the purpose of decoupling the system in Eq. (8.9) for the direct and independent symbolic integration of each individual equation.

The essence of modal analysis is the expression of the vector $y(t)$ of Eq. (8.9) in the form

$$y(t) = c_1(t)x_1 + c_2(t)x_2 + \cdots + c_N(t)x_N \qquad (8.10)$$

in which $x_1, x_2, \ldots, x_N$ is the orthonormal eigenvector system of the (global stiffness) matrix K. Being orthonormal, the *fixed* eigenvectors span the N-dimensional vector space inside which $y(t)$ moves with time. To substitute $y(t)$ into $Ky + \dot{y} = 0$, we need

$$\dot{y} = \dot{c}_1 x_1 + \dot{c}_2 x_2 + \cdots + \dot{c}_N x_N \qquad (8.11)$$

and have that

$$(\dot{c}_1 + \lambda_1 c_1)x_1 + (\dot{c}_2 + \lambda_2 c_2)x_2 + \cdots + (\dot{c}_N + \lambda_N c_N)x_N = 0 \qquad (8.12)$$

in which λ_j is from $Kx_j = \lambda_j x_j$. Premultiplying Eq. (8.12) successively by $x_1^T, x_2^T, \ldots, x_N^T$ produces the decoupled system

$$\dot{c}_1 + \lambda_1 c_1 = 0$$
$$\dot{c}_2 + \lambda_2 c_2 = 0$$
$$\vdots$$
$$\dot{c}_N + \lambda_N c_N = 0$$

(8.13)

for $c_1, c_2, \ldots, c_N$. Each of Eqs. (8.13) can now be solved independently in the symbolic form

$$c_j(t) = c_j(0)e^{-\lambda_j t}$$

(8.14)

and according to Eq. (8.10)

$$y(t) = \sum_{j=1}^{N} c_j(0)x_j e^{-\lambda_j t}$$

(8.15)

in which the initial values $c_j(0)$ are obtained from Eq. (8.10) by

$$c_j(0) = x_j^T y(0)$$

(8.16)

where $y(0)$ is the initial temperature distribution vector.

We know the global stiffness matrix to be generally positive semidefinite with positive eigenvalues. For the particular K in Eq. (8.8), we even know that $\lambda_j \cong j^2 \pi^2/\alpha$, hence all terms in Eq. (8.15) decrease with time, and the higher j is, the faster $e^{-\lambda_j t}$ decreases. As time elapses, the temperature distribution evens, and the importance of the higher modes x_j in $y(t)$ diminish faster than the lower ones, leaving the temperature distribution in the long run to be described principally by the modes at the lower end of the eigenvalues spectrum.

Modal or Fourier solution of the system $Ky + \dot{y} = 0$ is appropriate and attractive only when the eigenvectors and eigenvalues of K are to be had at a reasonable price, or when the temperature distribution is smooth to start with, necessitating only a few of the lower modes for the adequate description of $y(t)$, for then the eigensystem of K is computed once and $y(t)$ is followed to any time length by Eq. (8.15). Otherwise, if K is large and of a form that requires expensive computations to extract its eigensystem, a finite difference discretization in time for the step by step integration of $Ky + \dot{y} = 0$, starting with the initial $y_0 = y(0)$, is more desirable. Before we investigate the large subject of these time integration algorithms for $Ky + \dot{y} = 0$, we shall consider the finite element techniques for reducing the first-order boundary value problem into this form.

4. Finite Elements

When we look at Eq. (8.4) at a fixed moment of time, consider $\dot{y}$ a given load or source function and resolve to first discretize it only in space; finite elements may be applied for this purpose as routinely as in boundary value problems. To simplify the writing, we put Eq. (8.4) in the form

$$-(p(x)y')' + q(x)y = -\rho(x)\dot{y} \tag{8.17}$$

and assume that the ends are either at a prescribed temperature, y, or insulated so that $y' = 0$. Considering $\dot{y}$ given, we write for Eq. (8.17) the total potential energy

$$\pi(y) = \tfrac{1}{2}\int_0^1 [p(x)y'^2 + q(x)y^2 + 2\rho(x)y\dot{y}]\,dx \tag{8.18}$$

precisely as in Eq. (3.20), except that $f(x)$ is replaced here by $-\rho(x)\dot{y}$. To discretize $\pi(y)$ with finite elements, we take both y and $\dot{y}$ as nodal values, interpolate them inside the element with the *same* polynomial shape functions, and have for the typical eth element that

$$\pi_e = \tfrac{1}{2}y_e{}^{\mathrm{T}}k_e y_e + y_e{}^{\mathrm{T}}m_e \dot{y}_e \tag{8.19}$$

where

$$(k_e)_{ij} = \int_{x_1}^{x_1+h} [p(x)\phi_i'\phi_j' + q(x)\phi_i\phi_j]\,dx \tag{8.20}$$

and

$$(m_e)_{ij} = \int_{x_1}^{x_1+h} \rho(x)\phi_i\phi_j\,dx \tag{8.21}$$

and where y_e and $\dot{y}_e$ include the element nodal values for y and $\dot{y}$, respectively. Assembly of π_e into π, as described in Chapter 4, Section 4, results in the global expression

$$\pi = \sum_{e=1}^{Ne} \pi_e = \tfrac{1}{2}y^{\mathrm{T}}Ky + y^{\mathrm{T}}M\dot{y} \tag{8.22}$$

into which the essential boundary conditions still need to be introduced.

5. Essential Boundary Conditions

At an insulated end, $y' = 0$. This is a natural boundary condition, included in the approximation process and which we do not have to enforce explicitly; as the mesh is refined to improve the accuracy of the computed solution, this boundary condition is also better approximated. If both ends of the rod are insulated, no constraints are imposed on π in Eq. (8.22) and $\partial\pi/\partial y = 0$ produces the system of equations

$$Ky + M\dot{y} = 0 \qquad (8.23)$$

similar to that in Eq. (8.9), obtained with finite differences, except that here M is nondiagonal and K only positive *semi*definite.

In case the ends are held at some prescribed temperature, this is an essential boundary condition which y, in Eq. (8.22), must satisfy at all times. We let the prescribed temperature y_1 be a given function of time and assume that at the end point, $x = 0$:

$$y_1 = y(0, t) = y_1(t), \qquad \dot{y}_1 = \dot{y}(0, t) = \dot{y}_1(t) \qquad (8.24)$$

Since at any moment of time y_1 is fixed in space, $\partial\pi/\partial y_1 = 0$ in Eq. (8.22), and $\partial\pi/\partial y = 0$ produces the system that we explicitly write as

$$\begin{bmatrix} K_{21} & \cdots & K_{2N} \\ \vdots & & \vdots \\ K_{N1} & \cdots & K_{NN} \end{bmatrix} \begin{bmatrix} y_1 \\ \vdots \\ y_2 \\ \vdots \\ y_N \end{bmatrix} + \begin{bmatrix} M_{21} & \cdots & M_{2N} \\ \vdots & & \vdots \\ M_{N1} & \cdots & M_{NN} \end{bmatrix} \begin{bmatrix} \dot{y}_1 \\ \vdots \\ \dot{y}_2 \\ \vdots \\ \dot{y}_N \end{bmatrix} = 0 \qquad (8.25)$$

or concisely after partitioning

$$K_1 y_1 + Ky + M_1 \dot{y}_1 + M\dot{y} = 0 \qquad (8.26)$$

in which $K_1^{\mathrm{T}} = (K_{21}, \ldots, K_{N1})$, $M_1^{\mathrm{T}} = (M_{21}, \ldots, M_{N1})$, $y^{\mathrm{T}} = (y_2, \ldots, y_N)$, and K and M are the same as in Eq. (8.23), but with their first row and column taken out. Moving the given vectors in Eq. (8.26) to the right brings this non-homogeneous equation to the form

$$Ky + M\dot{y} = -K_1 y_1 - M_1 \dot{y}_1 = f(t) \qquad (8.27)$$

We have already mentioned the programmer's inconvenience in changing the size of two-dimensional arrays. To avoid this change and retain K and M in their original size, prior to the introduction of the essential boundary conditions, the first row and column is restored to Eq. (8.27) by adding to it the prescribed temperature in the form

$$\begin{bmatrix} 0 & 0 \\ \hline 0 & K \end{bmatrix} \begin{bmatrix} y_1 \\ y_2 \\ \vdots \\ y_N \end{bmatrix} + \begin{bmatrix} 1 & 0 \\ \hline 0 & M \end{bmatrix} \begin{bmatrix} \dot{y}_1 \\ \dot{y}_2 \\ \vdots \\ \dot{y}_N \end{bmatrix} = \begin{bmatrix} \dot{y}_1 \\ f \end{bmatrix} \qquad (8.28)$$

which is designed to render M nonsingular. Instead, we could have made K invertible by adding the prescribed y_1 to Eq. (8.28) in place of $\dot{y}_1$: This is achieved by putting a 1 at the upper left-hand corner of K, a zero at the same place in M, and y_1 at the first entry of the right-hand side vector in Eq. (8.28) instead of $\dot{y}_1$.

To apply the modal analysis of Section 3 to the more general nonhomogeneous system $Ky + M\dot{y} = f(t)$, we stand in need of the eigensystem of the general eigenproblem $Kx = \lambda Mx$. Once more y and $\dot{y}$ are expanded in terms of the eigenvectors in the form $y = c_1 x_1 + c_2 x_2 + \cdots + c_N x_N$, $\dot{y} = \dot{c}_1 x_1 + \dot{c}_2 x_2 + \cdots + \dot{c}_N x_N$, and $Ky + M\dot{y} = f(t)$ becomes

$$\sum_{j=1}^{N} c_j K x_j + \sum_{j=1}^{N} \dot{c}_j M x_j = f(t) \tag{8.29}$$

or with $Kx_j = \lambda_j M x_j$,

$$\sum_{j=1}^{N} (\lambda_j c_j + \dot{c}_j) M x_j = f(t) \tag{8.30}$$

Inasmuch as the eigenvectors form an orthonormal system with respect to M, $x_i^T M x_j = 0$ when $i \neq j$ and $x_j^T M x_j = 1$, the premultiplication of Eq. (8.30) by $x_1^T, x_2^T, \ldots, x_N^T$ decouples it into

$$\lambda_j c_j + \dot{c}_j = x_j^T f \tag{8.31}$$

which are solved one by one.

6. Euler's Stepwise Integration in Time

If not by modal analysis, the system $Ky + M\dot{y} = f$ may be solved, if it is further discretized by finite differences and y pursued in time, step by step, starting with the initial temperature distribution $y_0 = y(t = 0)$. Euler's method is one of the earliest and simplest of these time marching schemes and is based on a *backward* difference formula for $\dot{y}$. Let y_j be the value of the temperature *vector* at time t, and y_{j+1} the same at time $t + \tau$. Then approximately, $\dot{y}_j = (y_{j+1} - y_j)/\tau$, and at time t the equation $Ky + \dot{y} = 0$ is replaced by

$$Ky_j = -(1/\tau)(y_{j+1} - y_j) \tag{8.32}$$

or

$$y_{j+1} = (I - \tau K)y_j \tag{8.33}$$

with which we predict the temperature distribution at time $t + \tau$ from that at t. Starting with y_0 at $t = 0$, we successively compute

$$y_1 = (I - \tau K)y_0$$

$$y_2 = (I - \tau K)y_1 = (I - \tau K)^2 y_0$$

$$\vdots \tag{8.34}$$

$$y_n = Z^n y_0, \qquad Z = I - \tau K$$

where Z is the *magnification matrix* of Euler's scheme.

A decisive trait of Euler's method, responsible for making it a candidate for serious practical applications, is that it is *explicit*; the passage from time level t to time level $t + \tau$ does not involve the cumbersome inversion of a matrix nor the solution of a system of equations. It merely necessitates the multiplication of a vector by a matrix, an operation that can be compactly and efficiently carried out with finite element matrices, as explained in the next section. However, Euler's method ceases to be explicit when applied to $Ky + M\dot{y} = f$ with a nondiagonal M, typically generated by finite elements, since in this case the time scheme is of the form

$$My_{j+1} = (M - \tau K)y_j + \tau f_j \tag{8.35}$$

and a system of linear equations must be solved in order to obtain y_{j+1} from y_j. Matters are considerably simplified when M is diagonalized or lumped as was done in Chapter 6, Section 10, for the eigenproblem.

Euler's scheme for $Ky + \dot{y} = 0$ in Eq. (8.33) can be interpreted as the truncated analytic solution of this equation with y_j as initial condition. Indeed, the symbolic solution of $Ky + \dot{y} = 0$ is

$$y = e^{-Kt} y_0 \tag{8.36}$$

or in the interval $t, t + \tau$

$$y_{j+1} = e^{-K\tau} y_j \tag{8.37}$$

But

$$e^{-\tau K} = I - \tau K + \frac{1}{2!} \tau^2 K^2 - \frac{1}{3!} \tau^3 K^3 \pm \cdots \tag{8.38}$$

As $\tau \to 0$, we may approximately write

$$e^{-\tau K} = I - \tau K \tag{8.39}$$

and hence Eq. (8.33).

The expansion of $e^{-\tau K}$ in powers of τK suggests the quadratic, more accurate, Euler scheme

$$y_{j+1} = (I - \tau K + \tfrac{1}{2}\tau^2 K^2)y_j \tag{8.40}$$

In the practical application of Eq. (8.40), the explicit formation of K^2 is avoided because it would require the storage of an additional matrix K^2, which is also *denser* than the original stiffness matrix K. Instead, $K^2 y_j$ is formed by the successive operations $K(Ky_j)$.

7. Explicit Finite Element Schemes

The product Ky_j needed in Euler's (or for that matter any other explicit) scheme can generally be formed in the finite element method *without the actual assembly of the global stiffness matrix K.* To simplify the writing, we drop the subscript j from y_j and proceed to show how Ky can be formed directly from the element data. According to Eq. (4.30), we may write Ky as

$$Ky = \sum_{e=1}^{Ne} A_e^{T} k_e A_e y \tag{8.41}$$

and carry out the multiplication in the following five steps:

(i) An empty array of length N is reserved for Ky.

(ii) For the typical eth element, $y_e = A_e y$ is formed by picking out from y, according to the local–global numbering list, the entries corresponding to this particular element.

(iii) The element vector $k_e y_e$ is formed. If the element is of n nodal values, then this operation involves only $n \times n$ arrays.

(iv) The element vector $k_e y_e$ is changed from the local numbering system to the global by $A_e^{T}(k_e y_e)$.

(v) The last vector is added to the array reserved for Ky, and the procedure repeated over all the Ne finite elements. Notice, however, that in step (iv) $A_e^{T}(k_e y_e)$ need not be formed explicitly; the entries of $k_e y_e$ have only to be properly added to Ky.

In the event of a uniform mesh only *one single element* need be stored, in addition to the local–global numbering list, in order to completely specify K and for Ky to be carried out. Also, the formation of Ky by Eq. (8.41), as detailed in the above steps (i)–(v), need not take into account the essential boundary conditions that can be introduced *after* Ky has been created. For instance, if y_1—the first entry of y—is equal to zero at all times, then, as explained in Section 5, the first row and column of K are set equal to zero or equivalently, y_1 is set to zero, Ky formed, and then also $(Ky)_1$ is set equal to zero.

8. Convergence of Euler's Method

Convergence of the space discretization has been amply treated in Chapter 5. Presently, we want to establish the convergence of the time discretization, using Euler's scheme. Since the coupled system of equations $Ky + \dot{y} = 0$ that we are interested in can be decoupled, we restrict our analysis to the first-order initial value problem

$$\dot{y} + \lambda y = 0, \qquad \lambda > 0, \qquad t > 0, \qquad y(0) = y_0 > 0 \qquad (8.42)$$

with only one unknown function $y = y(t)$. Much of what is particular to the convergence of step by step integration schemes will surface during this deliberately simplified study. In computing the error in the temperature, we shall naturally need to distinguish between the *computed* value which we denote by y_j and the *exact* value which we denote by Y_j. Now, we have from Taylor's theorem that between time levels j and $j + 1$

$$Y_{j+1} = Y_j + \tau \dot{Y}_j + \tfrac{1}{2}\tau^2 \ddot{Y}(\theta) \qquad (8.43)$$

where θ is somewhere between time level j and $j + 1$. Because Y satisfies Eq. (8.42), $\dot{Y}_j = -\lambda Y_j$, $\ddot{Y}(\theta) = \lambda^2 Y(\theta)$ and Eq. (8.43) becomes

$$Y_{j+1} = zY_j + \tfrac{1}{2}\tau^2\lambda^2 Y(\theta), \qquad z = 1 - \lambda\tau \qquad (8.44)$$

Also, since Y is monotonically decreasing,

$$Y_{j+1} \leqslant Y(\theta) \leqslant Y_j \qquad (8.45)$$

For brevity we introduce the notation

$$\zeta_j = \tfrac{1}{2}(Y(\theta)/Y_j)\lambda^2\tau^2 \leqslant \tfrac{1}{2}\lambda^2\tau^2 \qquad (8.46)$$

using which, Eq. (8.44) is written as

$$Y_{j+1} = zY_j + \zeta_j Y_j \qquad (8.47)$$

In view of the fact that

$$y_{j+1} = zy_j, \qquad z = 1 - \lambda\tau \qquad (8.48)$$

the term $\zeta_j Y_j$ in Eq. (8.47) is the residual of Euler's scheme and is here $O(\lambda^2 \tau^2)$. Substraction of Eq. (8.48) from Eq. (8.47) produces the recursive error formula

$$e_{j+1} = ze_j + \zeta_j Y_j \qquad (8.49)$$

where $e_j = Y_j - y_j$. Following the error from $e_0 = Y_0 - y_0 = 0$ to e_n, we find that

$$e_n = \zeta_0 z^{n-1} Y_0 + \zeta_1 z^{n-2} Y_1 + \cdots + \zeta_{n-1} Y_{n-1} \qquad (8.50)$$

As $\tau \to 0$, a stage is reached for which $0 < z < 1$; then all terms in Eq. (8.50) are positive, and since $\zeta_j \leqslant \frac{1}{2}\lambda^2\tau^2$, e_n becomes

$$e_n \leqslant \frac{1}{2}\lambda^2\tau^2(z^{n-1} Y_0 + z^{n-2} Y_1 + \cdots + Y_{n-1}) \qquad (8.51)$$

Furthermore, since $\zeta_j Y_j \geqslant 0$, Eq. (8.47) asserts that $Y_{j+1} \geqslant z Y_j$, so that e_n in Eq. (8.51) becomes

$$e_n \leqslant \frac{1}{2}\lambda^2\tau^2 n Y_{n-1} \qquad (8.52)$$

and since $n\tau = t$ and $Y_{n-1} = Y_0 \exp[-\lambda(n-1)\tau] \cong Y_0 \exp(-\lambda t)$, we finally have our error bound

$$e_n \leqslant \frac{1}{2} Y_0 \lambda^2 \tau t e^{-\lambda t} \qquad (8.53)$$

and for any *fixed* time t, $e_n \to 0$ as $\tau \to 0$. The *truncation* error in Euler's scheme is $O(\tau^2)$, but after these errors have been summed up over n time steps, each of size τ, the error $e_n = Y_n - y_n$ in y_n is left at only $O(\tau)$.

The truncation error in the quadratic Euler scheme in Eq. (8.40) is $O(\tau^3)$ and the present error analysis predicts for it that $e_n = O(\tau^2)$, meaning that for a comparable accuracy fewer steps or a longer τ is needed with the scheme $y_{j+1} = (1 - \lambda\tau + \frac{1}{2}\lambda^2\tau^2)y_j$ than with $y_{j+1} = (1 - \lambda\tau)y_j$. But as we shall soon discover for ourselves in the next section, *stability* considerations drastically limit the allowable step size in both schemes, injecting thereby a wholly new argument into the question of step size selection.

9. Stability

While in convergence analysis we considered y_n at a *fixed* time and strived to discover the way in which $e_n = Y_n - y_n$ diminishes as t is reached with an even larger number of smaller time intervals τ, in stability analysis we consider y_n for a *fixed* time step τ and observe the behavior of $e_n = Y_n - y_n$ as n increases.

As far as the stability analysis of Euler's method is concerned, we wish first to look at it for the single equation $\dot{y} + \lambda y = 0$, $y(0) = y_0$, with $\lambda > 0$, whose exact solution is $y = y_0 e^{-\lambda t}$, and as $t \to \infty$, $y(t) \to 0$. With Euler's scheme $y_{j+1} = (1 - \lambda \tau) y_j$, $y_n = (1 - \lambda \tau)^n y_0$, and if $|1 - \lambda \tau| < 1$ also $y_n \to 0$ as $n \to \infty$, *but not* if $|1 - \lambda \tau| \geqslant 1$. When $|1 - \lambda \tau| > 1$, y_n oscillates with n, with an ever increasing amplitude, producing senseless results and a certain early abortion of the program due to overflow. Thus, Euler's method is only *conditionally stable*: It is *stable* when $|1 - \lambda \tau| < 1$ and *unstable* when $|1 - \lambda \tau| > 1$. To ensure the stable evolution of Euler's scheme, the time step τ has to be limited to

$$0 < \tau < 2/\lambda \qquad (8.54)$$

Stability becomes a serious problem with Euler's method in its application to whole systems of equations formed by finite differences or finite elements in the space discretization of the heat conduction problem. Let $Ky + M\dot{y} = 0$, $y(0) = y_0$, be with a positive definite and symmetric stiffness matrix K and mass matrix M. Here, Euler's method becomes

$$y_n = (I - \tau M^{-1} K)^n y_0 \qquad (8.55)$$

and in order to study the stability of this scheme we expand the initial vector y_0 in terms of the eigenvectors $x_1, x_2, \ldots, x_N$ of $Kx = \lambda Mx$ in the form $y_0 = c_1 x_1 + c_2 x_2 + \cdots + c_N x_N$, and have upon its introduction into Eq. (8.55) that

$$y_n = c_1 (1 - \lambda_1 \tau)^n x_1 + c_2 (1 - \lambda_2 \tau)^n x_2 + \cdots + c_N (1 - \lambda_N \tau)^n x_N \qquad (8.56)$$

In view of the fact that K and M are both positive definite, all the eigenvalues $\lambda_1 \leqslant \lambda_2 \leqslant \cdots \leqslant \lambda_N$ are positive, and for $y_n \to 0$ as $n \to \infty$, that is, for stability, we must have that $|1 - \lambda_j \tau| < 1$ for *all* $j = 1, 2, \ldots, N$. Now stability restricts τ to

$$0 < \tau < 2/\lambda_N \qquad (8.57)$$

and we know that λ_N is proportional to N^2.

To see how stability considerations override those of accuracy, consider the case of $y_0 = c_1 x_1$ where x_1 is the first eigenvector of $Kx_1 = \lambda_1 Mx_1$. Theoretically, $c_2, c_3, \ldots, c_N$ are absent from y_n and the stability restriction here is only $0 < \tau < 2/\lambda_1$. But round-off errors introduce spurious $c_2 \neq 0$, $c_3 \neq 0, \ldots, c_N \neq 0$ into Eq. (8.56) and these terms inevitably explode if for any j, $|1 - \lambda_j \tau| > 1$. Thus, even if $y_0 = c_1 x_1$ only, we are still restricted to $\tau < 2/\lambda_N$ and are forced to march in much smaller time steps than are actually needed for this y_0.

For the quadratic Euler scheme in Eq. (8.40), stability requires that

$$|z| = |1 - \lambda \tau + \tfrac{1}{2} \lambda^2 \tau^2| < 1 \qquad (8.58)$$

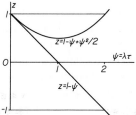

Fig. 8.2 Variation with $\psi = \lambda\tau$ of the magnification factors z of the linear and quadratic Euler schemes.

The variation of z with $\psi = \lambda\tau$ is graphically shown in Fig. 8.2 for both this and the linear scheme. We learn from this figure that for the quadratic scheme the stable τ is in the range $0 < \tau < 2/\lambda_N$. Conclusion: Wherever τ is restricted by stability considerations, the use of the more accurate quadratic Euler scheme is wasteful.

10.   Stable Time Step Size Estimate

In order not to surpass the stable time step size in Euler's scheme, or actually in any other conditionally stable time scheme, we need an a priori estimate of the largest eigenvalue λ_N of the general eigensystem $Ky = \lambda My$. This is conveniently and realistically provided by the analysis of Chapter 7, Section 1, in which we found that

$$\lambda_N \leqslant \max_e\{\lambda_n{}^e\}, \qquad e = 1, 2, \ldots, Ne \tag{8.59}$$

where $\lambda_n{}^e$ is the largest eigenvalue of the element system

$$k_e y_e = \lambda^e m_e y_e, \qquad e = 1, 2, \ldots, N \tag{8.60}$$

As an example of this estimate, consider the heat conduction Eq. (8.6) discretized with the quadratic element matrices

$$k_e = \frac{1}{6\alpha h}\begin{bmatrix} 7 & -8 & 1 \\ -8 & 16 & -8 \\ 1 & -8 & 7 \end{bmatrix}, \qquad m_e = \frac{h}{15}\begin{bmatrix} 4 & 2 & -1 \\ 2 & 16 & 2 \\ -1 & 2 & 4 \end{bmatrix} \tag{8.61}$$

It is a simple exercise to evaluate the eigenvalues of $k_e y_e = \lambda^e m_e y_e$ and we find that the largest is $\lambda_3{}^e = 15/(\alpha h^2)$. Hence, according to Eq. (8.59), $\lambda_N \leqslant 15/(\alpha h^2)$ and the stable time step τ is, from Eq. (8.57),

$$0 < \tau < \tfrac{2}{15}\alpha h^2 \tag{8.62}$$

In general, the upper limit on τ is $O(h^2)$, independent of the degree of the polynomial shape functions, and the use of higher-order elements in the spatial discretization, which calls for fewer of them (or equivalently, a larger h) to achieve a certain accuracy, consequently makes a larger stable time step τ permissible.

The appeal of explicit time schemes for the system $Ky + \dot{y} = 0$ lies in the fact that progression in time with these schemes does not call for a matrix inversion, matrix decomposition with back substitution, or the solution of a large system of linear equations; only the multiplication of a vector by a matrix, which can be very compactly and efficiently done on the element data. But finite elements naturally produce the more general coupled system $Ky + M\dot{y} = 0$ and M^{-1} need be formed or a large system of equations solved in each step of Euler's method. The computational complications introduced by this are obvious, taking into account that M^{-1} is dense even though M is sparse. To benefit from the advantages the method of finite elements offers to the practical programmer, retaining the use of Euler's method but avoiding the inversion of large matrices, we seek to lump M into a diagonal matrix.

Lumping also has the interesting side effect of extending stability. For the quadratic element, m_e is best lumped in the form

$$m_e = \frac{h}{3} \begin{bmatrix} 1 & & \\ & 4 & \\ & & 1 \end{bmatrix} \tag{8.63}$$

Now the largest λ^e in $k_e y_e = \lambda^e m_e y_e$, with k_e as in Eq. (8.61), is $\lambda_3{}^e = 6/(\alpha h^2)$ and the stability of Euler's method is assured with

$$0 < \tau < \tfrac{1}{3}\alpha h^2 \tag{8.64}$$

a predicted maximal time step, which is 2.5 times larger than that in Eq. (8.62), which was obtained from the *consistent* element mass matrix in Eq. (8.61). Numerical examples corroborate this prediction that lumping extends the range of the stable τ.

In our quest for stability, we are tempted to trade accuracy for its sake and write the diagonal mass matrix for the quadratic element in the form

$$m_e = h \begin{bmatrix} m & & \\ & 2(1 - m) & \\ & & m \end{bmatrix} \tag{8.65}$$

which assures a correct kinetic energy for a constant displacement y for any value of m. The two nonzero eigenvalues of $k_e y_e = \lambda^e m_e y_e$ with k_e as in Eq. (8.61) and m_e as in Eq. (8.65) are

$$\lambda_2{}^e = \frac{1}{\alpha m} h^{-2}, \qquad \lambda_3{}^e = \frac{4}{3\alpha m(m - 1)} h^{-2} \tag{8.66}$$

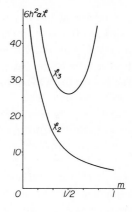

Fig. 8.3 Eigenvalues of $k_e y_e = \lambda^e m_e y_e$ with the element stiffness matrix k_e in Eq. (8.61) and the lumped element mass matrix in Eq. (8.65).

shown in Fig. 8.3 in their dependence upon m. From the stability viewpoint, the best m is seen (Fig. 8.3) to be $m = \frac{1}{2}$, for which $\lambda_3{}^e = 16/(3\alpha h^2)$ as compared with $\lambda_3{}^e = 6/(\alpha h^2)$ when $m = \frac{1}{3}$, which is the optimal m from the discretization viewpoint. No significant improvement in the stability is gained with $m = \frac{1}{2}$ and we reject it in favor of $m = \frac{1}{3}$.

11. Numerical Example

Loss of stability in the time schemes is sudden and violent, as illustrated in the following numerical example which involves the use of finite elements and then Euler's method in the computation of the temperature distribution and evolution in a heat conducting rod insulated along its entire length and also its end point at $x = 1$. We assume the rod to be at zero initial temperature except for the end point $x = 0$, which is instantaneously raised to the temperature $y(0, t) = 1$ at $t = 0$, and which is kept at this temperature always. We further assume the thermal diffusivity coefficient α of the rod to be one and formally write this initial value problem as

$$y'' = \dot{y}, \quad 0 < x < 1, \quad t > 0$$

$$y(0, t) = 1, \quad y'(1, t) = 0, \quad t > 0 \qquad (8.67)$$

$$y(x, 0) = 0, \quad 0 \leqslant x \leqslant 1$$

whose solution is known to be

$$y(x, t) = 1 - \sum_{j=0}^{\infty} \frac{4}{(2j - 1)\pi} \exp\left[-\frac{1}{4}\pi^2(2j - 1)^2 t\right] \sin\left[\frac{1}{2}\pi(2j - 1)x\right] \qquad (8.68)$$

Discretization in x of Eq. (8.67) is performed with seven quadratic finite elements ($h = \frac{1}{14}$) uniformly distributed, and with the lumped mass matrix in Eq. (8.63). After the global matrices K and M have been formed, we proceed to follow the change in the temperature with time using Euler's method. Eq. (8.64) tells us that for stability we need to have $\tau < \alpha h^2/3 = 0.0017$ in this case. The computed temperature is observed at the point $x = 4h$ and is traced as a function of time in Fig. 8.4. For the choice $\tau = 0.0018$, which is above the stable time step size predicted in Eq. (8.64), the computed temperature oscillates violently with ever increasing amplitude, as can be seen in Fig. 8.4a, and the scheme is unstable, resulting soon in the premature abortion of the program. It is sufficient that we reduce the time step to $\tau = 0.0015$ for stability to be restored to Euler's scheme as can be seen in Fig. 8.4b, demonstrating also the usefulness of the bound on λ_N, and eventually on τ, in Eq. (8.59).

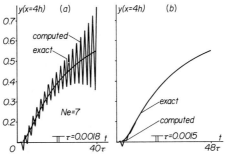

Fig. 8.4 Exact and computed (using Euler's scheme) temperature evolution at $x = 4h$ in the conducting rod problem in Eq. (8.67) discretized in space with seven quadratic finite elements.

The short lived spurious oscillations in the computed temperature observed in Fig. 8.4b can be explained by reference to Eq. (8.56). When τ is close to its stability limit, as inevitably it usually is for accuracy reasons, some of the terms $1 - \lambda_j\tau$ turn negative and when j is close to N, they are only slightly greater than -1. At the beginning, this causes $(1 - \lambda_j\tau)^n$ to oscillate between nearly -1 and 1 inducing in turn high frequency oscillations in the computed y_n. But since $|1 - \lambda_j\tau| < 1$, the amplitude of these oscillations diminishes with time to their eventual demise. How severe these oscillations are depends on $c_j, j \gg 1$, in Eq. (8.56). If c_N is relatively large due to discontinuities in the initial temperature, more serious short period oscillations are to be expected than when c_N is relatively small due to a smooth initial temperature distribution.

12. Implicit Unconditionally Stable Schemes

A slight modification of Euler's method turns it into an *implicit one step* time integration scheme which is *unconditionally stable*—stable for any positive time step τ. This scheme, habitually named after Crank–Nickolson, or otherwise called the trapezoidal rule, is of great use in our heat conduction problem that gives rise to a *stiff* system of equations $Ky + \lambda M\dot{y} = 0$, meaning that the eigenvalues of $Kx = \lambda Mx$ have a wide range from $O(1)$ to $O(h^{-2})$. For in this case, stability seriously restricts the size of τ in Euler's method, usually to a magnitude which is below that sufficient for the prospective accuracy.

Here, we expand y_1 in terms of y_0, $\dot{y}_0$, *and* $\dot{y}_1$ in the form

$$y_1 = y_0 + \tfrac{1}{2}\tau(\dot{y}_0 + \dot{y}_1) \tag{8.69}$$

which is correct for $y = 1$, $y = t$, and $y = t^2$, and obtain for the system $Ky + \dot{y} = 0$, with $\dot{y}_0 = -Ky_0$ and $\dot{y}_1 = -Ky_1$, the scheme

$$y_1 = (I + \tfrac{1}{2}\tau K)^{-1}(I - \tfrac{1}{2}\tau K)y_0 \tag{8.70}$$

which is implicit because of $(I + \tfrac{1}{2}\tau K)^{-1}$. The origin of this implicitness is of course, $\dot{y}_1$ to the right of Eq. (8.69). For the single equation $\dot{y} + \lambda y = 0$, the trapezoidal rule in Eqs. (8.70) becomes

$$y_1 = zy_0, \qquad z = (1 - \tfrac{1}{2}\psi)/(1 + \tfrac{1}{2}\psi), \qquad \psi = \lambda\tau \tag{8.71}$$

and the variation of z, its magnification factor, with $t = 0$ is shown in Fig. 8.5. It is equal to 1 at $\psi = 0$ and monotonically decreases to $z = -1$ as $\psi \to \infty$. In between, for any $0 < \psi < \infty$, $|z| < 1$, and the method is stable for all $\tau > 0$; thus, it is unconditionally stable. The price of this unlimited stability is implicitness, which might, under proper circumstances, be worth paying. The difference scheme in Eq. (8.71) is with a residual $O(\tau^3)$, such that at time $t > 0$, the error in y_n is $O(\tau^2)$, double that of Euler's, and without the

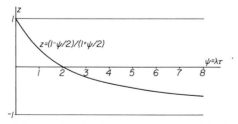

Fig. 8.5 Magnification factor for the trapezoidal rule.

restrictions of stability, a larger time step can actually be chosen in the trapezoidal rule.

To prove the unconditional stability of the trapezoidal scheme applied to $Ky + M\dot{y} = 0$, we replace K in Eq. (8.70) with $M^{-1}K$, and write y_n as

$$y_n = Z^n y_0, \qquad Z = (I + \tfrac{1}{2}\tau M^{-1}K)^{-1}(I - \tfrac{1}{2}\tau M^{-1}K) \qquad (8.72)$$

then expand y_0 in terms of the eigenvectors $x_1, x_2, \ldots, x_N$ of the eigenproblem $Kx = \lambda M x$ and get

$$y_n = \sum_{j=1}^{N} c_j \left(\frac{1 - \tfrac{1}{2}\lambda_j \tau}{1 + \tfrac{1}{2}\lambda_j \tau} \right)^n x_j \qquad (8.73)$$

Inasmuch as both K and M are positive definite, $\lambda_j > 0$ for all $j = 1, 2, \ldots, N$, and each of the magnification factors

$$z_j = (1 - \tfrac{1}{2}\psi_j)/(1 + \tfrac{1}{2}\psi_j), \qquad \psi_j = \lambda_j \tau \qquad (8.74)$$

is such that $|z_j| < 1$.

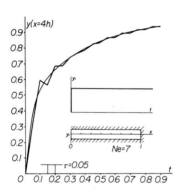

Fig. 8.6   Temperature evolution computed, using the trapezoidal rule, for the same heat conducting rod as in Fig. 8.4.

To observe numerically the performance of the implicit Crank–Nickolson scheme of Eq. (8.70), we apply it to the time integration of the heat conduction problem (8.67) discretized with the same seven quadratic finite elements. Taking advantage of the unconditional stability of this scheme, we boldly choose $\tau = 0.05$, which is some 33 times larger than the stable time step used to get the results in Fig. 8.4b, with Euler's method. Since matrix inversion is needed anyway in this implicit scheme, we forego the lumping of M and use, instead, the consistent element mass matrix in Eq. (8.61). Figure 8.6 shows the temperature thus computed at $x = 4h$ (between the second and third elements) as compared to the exact temperature.

The computed temperature is characterized here by high frequency oscillations that are not short lived, but that nevertheless diminish in amplitude with the passing of time. An explanation of this phenomenon is provided by Eqs. (8.73) and (8.74): Since we march, with this unconditionally stable scheme, in relatively large time steps τ, some of the magnification factors near $j = N$ become only slightly larger than -1, and z_j^n changes sign with each step, causing y_n to oscillate. However, because $|z_j| < 1$, these oscillations diminish in amplitude as time passes, and $|z_j|^n \to 0$ as $n \to \infty$. In this respect, these numerical oscillations are similar and have the same origin as those observed in Fig. 8.4b, except that here they linger much longer. When $\tau < 2/\lambda_N$, all z_j are positive and the oscillations in the computed temperature cease.

The swinging of the temperature computed using the trapezoidal rule above and below the exact value suggests the *averaging* of the computed temperature to enhance its accuracy. This can be done over one step by $y_a = (y_0 + y_1)/2$ or over two steps by $y_a = (y_0 + 2y_1 + y_2)/4$. For example, at $t = 0.25$, $t = 0.30$, and $t = 0.35$, we computed the values $y_0 = 0.68993$, $y_1 = 0.73973$, and $y_2 = 0.76935$. Averaging by $y_a = (y_0 + 2y_1 + y_2)/4$ yields $y_a = 0.735$, as compared with the exact $y_1 = 0.736$.

We further note with interest that the magnification factor z in Eq. (8.71) for the trapezoidal rule is the *rational* or the *Padé* approximation for $e^{-\psi}$

$$e^{-\psi} = (1 - \tfrac{1}{2}\psi)/(1 + \tfrac{1}{2}\psi) + O(\psi^3) \tag{8.75}$$

A higher-order rational approximation to $e^{-\psi}$ is given by

$$e^{-\psi} = (1 - \tfrac{1}{2}\psi + \tfrac{1}{12}\psi^2)/(1 + \tfrac{1}{2}\psi + \tfrac{1}{12}\psi^2) + O(\psi^5) \tag{8.76}$$

suggesting a higher-order Crank–Nickolson scheme. The pros and cons of such higher-order schemes applied to the system $Ky + M\dot{y} = 0$ are discussed next.

13. Higher-Order Single Step Implicit Schemes

A higher-order single step implicit scheme is set up by predicting y_1 in terms of $y_0, \dot{y}_0, \dot{y}_1, \ddot{y}_0$, and $\ddot{y}_1$ in the form

$$y_1 = y_0 + \tau(\alpha_0 \dot{y}_0 + \alpha_1 \dot{y}_1) + \tau^2(\beta_0 \ddot{y}_0 + \beta_1 \ddot{y}_1) \tag{8.77}$$

with $\alpha_0, \alpha_1, \beta_0$, and β_1 adjusted so that Eq. (8.77) is correct for the highest possible polynomial y. With four parameters at our disposal, we can make

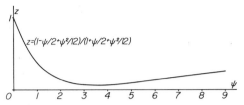

Fig. 8.7 Magnification factor of the implicit one step scheme in Eq. (8.79).

Eq. (8.77) correct for $y = t$, $y = t^2$, $y = t^3$, and $y = t^4$; the equation is already correct for $y = 1$. These conditions are met with $\alpha_0 = \alpha_1 = \frac{1}{2}$ and $\beta_0 = -\beta_1 = \frac{1}{12}$, and

$$y_1 = y_0 + \tfrac{1}{2}\tau(\dot{y}_0 + \dot{y}_1) + \tfrac{1}{12}\tau^2(\ddot{y}_0 - \ddot{y}_1) \tag{8.78}$$

For the single equation $\dot{y} + \lambda y = 0$, and with $\dot{y}_0 = -\lambda y_0$, $\dot{y}_1 = -\lambda y_1$, $\ddot{y}_0 = \lambda^2 y_0$, and $\ddot{y}_1 = \lambda^2 y_1$, Eq. (8.78) produces the scheme

$$y_1 = z y_0, \qquad z = \frac{1 - \tfrac{1}{2}\psi + \tfrac{1}{12}\psi^2}{1 + \tfrac{1}{2}\psi + \tfrac{1}{12}\psi^2}, \qquad \psi = \lambda \tau \tag{8.79}$$

as in Eq. (8.76). The change of this z with ψ is graphically shown in Fig. 8.7. The magnification factor for the scheme in Eq. (8.79) does not decrease monotonically as in the trapezoidal rule but, nevertheless, it is readily shown that for $\psi > 0$, $|z| < 1$, and this scheme is unconditionally stable. Also, whereas in the trapezoidal rule $z \to -1$ as $\psi \to \infty$, here $z > 0$ for all $\psi > 0$, and $z \to 1$ as $\psi \to \infty$.

Applied to the system $Ky + M\dot{y} = 0$, Eq. (8.77) produces

$$(M + \tfrac{1}{2}\tau K + \tfrac{1}{12}\tau^2 KM^{-1}K)y_1 = (M - \tfrac{1}{2}\tau K + \tfrac{1}{12}\tau^2 KM^{-1}K)y_0 \tag{8.80}$$

which can be solved for y_1 in terms of y_0 either by forming once and for all the magnification matrix

$$Z = (M + \tfrac{1}{2}\tau K + \tfrac{1}{12}\tau^2 KM^{-1}K)^{-1}(M - \tfrac{1}{2}\tau K + \tfrac{1}{12}\tau^2 KM^{-1}K) \tag{8.81}$$

and proceeding from time level 0 at t to time level 1 at $t + \tau$ by

$$y_1 = Z y_0 \tag{8.82}$$

or by factoring the matrix to the left of Eq. (8.80) in the form

$$U^T U = M + \tfrac{1}{2}\tau K + \tfrac{1}{12}\tau^2 KM^{-1}K \tag{8.83}$$

where U is upper triangular, and solving for each step the two systems

$$U^T x_1 = (M - \tfrac{1}{2}\tau K + \tfrac{1}{12}\tau^2 KM^{-1}K)y_0$$
$$U y_1 = x_1 \tag{8.84}$$

which requires only back substitution. When M is not lumped, $KM^{-1}K$ is dense and so is U, requiring vast computer storage. But even if M is diagonal, $KM^{-1}K$ is still considerably denser than K itself and of a much more complicated sparseness pattern, making the programming of this scheme considerably more complicated than that of the lower-order trapezoidal rule. In the next section we shall discuss a higher-order unconditionally stable scheme that overcomes this storage difficulty.

14. Superstable Schemes

The trapezoidal rule is unconditionally stable but as $\psi \to \infty$ in it, $z \to -1$, causing the computed temperature to oscillate rapidly, to remedy which we proposed averaging. *Superstable* methods have a magnification factor $z \to 0$ as $\psi \to \infty$ and such oscillations are absent when these schemes are applied to very stiff systems. The lowest-order super stable scheme is formed from the implicit prediction

$$y_1 = y_0 + \tau \dot{y}_1 \tag{8.85}$$

which is correct for $y = 1$ and $y = t$. For the single parameter equation $\dot{y} + \lambda y = 0$, it produces

$$y_1 = zy_0, \qquad z = 1/(1 + \psi), \qquad \psi = \lambda \tau \tag{8.86}$$

in which, indeed, $z \to 0$ as $\psi \to \infty$. It is typical for superstable schemes to have a rational magnification factor z with a higher power of ψ in the denominator than in the numerator. The variation of z with ψ in Eq. (8.86) is graphically shown in Fig. 8.8.

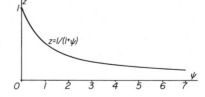

Fig. 8.8 Magnification factor of the super-stable scheme in Eq. (8.86).

For the system $Ky + M\dot{y} = 0$, the implicit prediction of Eq. (8.85) yields

$$(M + \tau K)y_1 = My_0 \tag{8.87}$$

which is unconditionally stable even if M is singular, provided that otherwise all eigenvalues of $Ky = \lambda My$ are positive. An additional advantage of the superstable scheme in Eq. (8.87) is that only one matrix $M + \tau K$ (comprising the two matrices M and K that have the same sparseness pattern) need be inverted or decomposed in it. But as is the case with implicit schemes, the handling of large two-dimensional arrays is unavoidable, and the scheme in Eq. (8.81) is also only of first order—its accuracy in time is no more than $O(\tau)$.

The creation of higher-order superstable schemes should be obvious and we do not wish to pursue this matter further. Computationally, the appearance of ψ^2 in the denominator of z in higher-order implicit schemes implies that the matrix $KM^{-1}K$ appears when this scheme is applied to $Ky + M\dot{y} = 0$, which, as we noted earlier, considerably complicates the programmer's task.

The higher-order scheme that we intend to present next is not exactly superstable, but on the other hand the denominator of its magnification factor consists of a *complete square* and consequently the inversion or decomposition of the quadratic matrix in Z can be successively carried out on the product of two *equal* matrices that include only K and M but not $KM^{-1}K$.

This practically useful scheme, credited to Calahan, is based on the approximation

$$y_1 = y_0 + \tau(\alpha_0 \dot{y}_0 + \alpha_1 \dot{y}_1) + \tau^2(\beta_0 \ddot{y}_0 + \beta_1 \ddot{y}_1) \qquad (8.88)$$

Requiring Eq. (8.88) to be correct for $y = 1$, $y = t$, $y = t^2$, and $y = t^3$ leaves us with still an extra parameter α and

$$y_1 = y_0 + \tau[\alpha \dot{y}_0 + (1 - \alpha)\dot{y}_1] + \tfrac{1}{6}\tau^2[(3\alpha - 1)\ddot{y}_0 + (3\alpha - 2)\ddot{y}_1] \quad (8.89)$$

which for the single parameter equation $\dot{y} + \lambda y = 0$, becomes the one step implicit scheme

$$[1 + \psi(1 - \alpha) - \tfrac{1}{6}\psi^2(3\alpha - 2)]y_1 = [1 - \psi\alpha + \tfrac{1}{6}\psi^2(3\alpha - 1)]y_0 \quad (8.90)$$

where, as usual, $\psi = \lambda\tau$. Had we decided to opt for $\alpha = \tfrac{1}{2}$, we would have obtained the scheme in Eq. (8.80). Instead, we select α so that the expression in the brackets multiplying y_1 in Eq. (8.90) is a complete square. That is, we look for a β, such that

$$(1 + \beta\psi)^2 = 1 + \psi(1 - \alpha) - \tfrac{1}{6}\psi^2(3\alpha - 2) \qquad (8.91)$$

and find that this occurs with

$$\alpha = \pm\sqrt{3}/3, \qquad \beta = \tfrac{1}{2}(1 - \alpha) \qquad (8.92)$$

Presently,

$$y_1 = zy_0, \qquad z = \frac{1 - \alpha\psi + \tfrac{1}{6}(3\alpha - 1)\psi^2}{[1 + \tfrac{1}{2}(1 - \alpha)\psi]^2} \qquad (8.93)$$

To choose between $\alpha = \sqrt{3}/3$ and $\alpha = -\sqrt{3}/3$, we start by observing that since at $\psi = 0$, $z = 1$, a necessary condition for the stability of the scheme in Eq. (8.93) is that

$$\left(\frac{dz}{d\psi}\right)_{\psi=0} < 0 \qquad (8.94)$$

In fact

$$\left(\frac{dz}{d\psi}\right)_{\psi=0} = -1 \qquad (8.95)$$

in this case independently of α and the scheme is at least conditionally stable. At $\psi = \infty$, we have from Eq. (8.93) that

$$z(\infty) = \tfrac{2}{3}(3\alpha - 1)/(1 - \alpha)^2 \qquad (8.96)$$

such that $z(\infty) = 2.732$ when α is positive, and $z(\infty) = -0.372$ when α is negative, and if the scheme is unconditionally stable it is so only with negative $\alpha = -\sqrt{3}/3$. To prove that the scheme in Eq. (8.93) is indeed unconditionally stable, we argue that in view of Eq. (8.95) and the fact that z depends continuously upon ψ, it is enough that $z \neq 1$ and $z \neq -1$ in the interval $0 < \psi < \infty$ for $|z|$ in this interval to be less than 1. Indeed, $z = 1$ and $z = -1$ when

$$-1 = \tfrac{1}{6}\psi(3 + 2\sqrt{3}) \qquad \text{and} \qquad \tfrac{1}{12}\psi^2(1 + 3\alpha^2) + 2\psi(\tfrac{1}{2} - \alpha) + 2 = 0 \qquad (8.97)$$

respectively, and when $\alpha = -\sqrt{3}/3$, neither of the equations in Eqs. (8.97) has a solution in the range $0 < \psi < \infty$. The variation of z with ψ in Eq. (8.93) is graphically shown in Fig. 8.9, where, as ψ tends to ∞ z tends to -0.732, and even though this scheme is not superstable in the sense that $z(\infty) = 0$, nevertheless, because $|z(\infty)| < 1$, we expect the solution computed with Calahan's scheme to be less oscillatory than that computed with the Crank–Nickolson method.

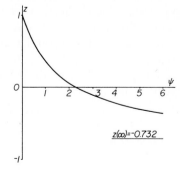

Fig. 8.9 Magnification factor of Calahan's scheme in Eq. (8.93) with $\alpha = -\sqrt{3}/3$.

The advantage of having a complete square in the donominator of z becomes manifest when Calahan's scheme is applied to the system $Ky + M\dot{y} = 0$. Suppose that M is lumped, such that

$$(M^{-1/2})_{ii} = (M_{ii})^{-1/2}, \qquad M^{-1/2}MM^{-1/2} = I \qquad (8.98)$$

We transform y and $\dot{y}$ according to

$$y = M^{-1/2}x, \qquad \dot{y} = M^{-1/2}\dot{x} \qquad (8.99)$$

and have that

$$M^{-1/2}KM^{-1/2}x + \dot{x} = 0 \qquad (8.100)$$

or in short

$$K'x + \dot{x} = 0, \qquad K' = M^{-1/2}KM^{-1/2} \qquad (8.101)$$

for which Eq. (8.90) takes the form

$$[(I + \beta\tau K')(I + \tau\beta K')]x_1 = [I - \alpha\tau K' + \tfrac{1}{6}\tau^2(3\alpha - 1)K'K']x_0 \qquad (8.102)$$

where $\alpha = -\sqrt{3}/3$ and $\beta = (1 - \alpha)/2$, and in which only *one* matrix $I + \tau\beta K'$ need be inverted or decomposed in order to pass from x_0 to x_1, and this matrix has the same sparseness pattern as the original K.

15. Multistep Schemes

In the previous sections we considered *single step* methods of time integration; methods that involve y and its time derivatives only at the beginning and the end of *one* time interval of size τ. Here we have in mind to extend this discussion to integration schemes that stretch over several time steps. The chief motivation for doing that is to create high-order schemes that do not involve higher-order derivatives of y since the time differentiation of y implies raising the power of K, which is patently undesirable.

First, we shall set up and analyze a two step method over time levels 0, 1, and 2 at t, $t + \tau$, and $t + 2\tau$, based upon the central difference formula $\dot{y}_1 = (y_2 - y_0)/2\tau$. For the single variable equation $\dot{y} + \lambda y = 0$, written at point 1, this leads to

$$(1/2\tau)(y_2 - y_0) + \lambda y_1 = 0 \qquad (8.103)$$

or, in general

$$y_{j+2} = y_j - 2\psi y_{j+1}, \qquad \psi = \lambda\tau \qquad (8.104)$$

Since there is only *one* initial condition $y(0) = y_0$ to the problem, but *two* values y_j and y_{j+1} are needed to predict y_{j+2}, the scheme in Eq. (8.104) is not *self-starting*. Another method, say Euler's, has to be used in order to compute y_1 from y_0 and only then can $y_3, y_4, \ldots$ be computed recursively with Eq. (8.104).

To study the stability of Eq. (8.104), we observe that the difference Eq. (8.104) admits the solution $y_j = cz^j$, which upon substitution into Eq. (8.104), produces the characteristic (auxiliary) equation

$$z^2 + 2\psi z - 1 = 0 \tag{8.105}$$

which posesses two roots z_1 and z_2; if they are distinct

$$y_j = c_1 z_1{}^j + c_2 z_2{}^j \tag{8.106}$$

while if $z_1 = z_2 = z$,

$$y_j = (c_1 + c_2 j)z^j \tag{8.107}$$

where c_1 and c_2 are determined by the (in this case) *two* initial conditions $y(0) = y_0$ and $y(\tau) = y_1$. From Eq. (8.106), we conclude that for stability independently of the initial conditions and hence c_1 and c_2, we have to have

$$|z_1| < 1, \qquad |z_2| < 1 \tag{8.108}$$

where $|z|$ stands for the modulus of z since it may well be complex. In the particular case of Eq. (8.105),

$$z_1 = -\psi + \sqrt{1 + \psi^2}, \qquad z_2 = -\psi - \sqrt{1 + \psi^2} \tag{8.109}$$

Both roots are real but $|z_2| > 1$ for all $\psi > 0$ and the method that appears so logical is *unconditionally unstable*, or simply unstable. In other words it is useless since in general $c_2 \neq 0$.

In our search for a stable, at least conditionally, explicit, two step method that does not include $\ddot{y}$, we write

$$y_2 = \alpha_0 y_0 + \alpha_1 y_1 + \tau(\beta_0 \dot{y}_0 + \beta_1 \dot{y}_1) \tag{8.110}$$

and adjust α_0, α_1, β_0, and β_1 so that Eq. (8.110) is exact for a polynomial of highest degree, while still leaving one parameter free to control stability. The conditions that y in Eq. (8.110) be exact for $y = 1$, $y = t$, and $y = t^2$ produce the system

$$\alpha_0 + \alpha_1 = 1$$
$$\alpha_1 + \beta_0 + \beta_1 = 2 \tag{8.111}$$
$$\alpha_1 + 2\beta_1 = 4$$

and for $\dot{y} + \lambda y = 0$, we have from this and Eq. (8.110) that

$$y_2 = [1 - \alpha - \tfrac{1}{2}\psi(3 + \alpha)]y_1 + [\alpha - \tfrac{1}{2}\psi(\alpha - 1)]y_0 \qquad (8.112)$$

in which α is the parameter left among α_0, α, β_0, and β_1. For the scheme in Eq. (8.110) to also be exact for $y = t^3$, we would have needed $\alpha = 5$.

To select α for stability, we write the characteristic equation for the scheme in Eq. (8.112):

$$z^2 - 2bz - c = 0, \qquad b = \tfrac{1}{2}(1 - \alpha) - \tfrac{1}{4}\psi(3 + \alpha), \qquad c = \alpha - \tfrac{1}{2}\psi(\alpha - 1) \quad (8.113)$$

which, when $\psi = 0$, possesses the two roots

$$z_1(\psi = 0) = 1, \qquad z_2(\psi = 0) = -\alpha \qquad (8.114)$$

and it is *necessary* for stability that

$$|\alpha| < 1, \qquad -1 < \alpha < 1 \qquad (8.115)$$

The discriminant Δ of Eq. (8.113) is

$$\Delta = \tfrac{1}{4}[\tfrac{1}{4}\psi^2(3 + \alpha)^2 + \psi(\alpha^2 - 1) + (1 + \alpha)^2] \qquad (8.116)$$

and is positive for any $|\alpha| < 1$. Consequently, the two roots z_1 and z_2 of Eq. (8.113) are real for $\psi > 0$. These roots depend continuously upon ψ, and to find the range of ψ inside which $|z| < 1$, it is enough that we set $z = 1$ and $z = -1$ into Eq. (8.113) and find the corresponding values of ψ. Doing that, we establish the limits of stability

$$0 < \psi < 1 - \alpha \qquad (8.117)$$

on ψ, or

$$0 < \tau < (1 - \alpha)/\lambda \qquad (8.118)$$

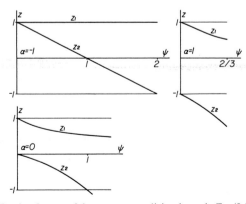

Fig. 8.10   Magnification factors of the two step explicit scheme in Eq. (8.112) for different αs.

on τ. If we choose $\alpha = -1$ in order to maximize the stability range, we get that $z_1 = 1$ and $z_2 = 1 - \psi$. Because $z_1 = 1$, this scheme does not lead to exactly $y_\infty = 0$, but rather to $y_\infty = c_1$. The variation of z_1 and z_2 from Eq. (8.113) with ψ is shown in Fig. 8.10 for different values of α.

When $\alpha = 0$, the scheme in Eq. (8.112) becomes

$$y_2 = \tfrac{1}{2}\psi y_0 + (1 - \tfrac{3}{2}\psi)y_1 \qquad (8.119)$$

which is explicit, conditionally stable in the range $0 < \tau < 1/\lambda$, and happens to be one of the *Adams–Bashforth* formulas. Still with $\alpha = 0$, if ψ is small, such that ψ^2 may be neglected relative to it, then the two roots of the characteristic equation (8.113) become

$$z_1 = 1 - \psi, \qquad z_2 = -\tfrac{1}{2}\psi \qquad (8.120)$$

and

$$y_j = c_1(1 - \psi)^j + c_2(-\tfrac{1}{2}\psi)^j \qquad (8.121)$$

Between the two roots of Eq. (8.113), $z_1{}^j$ approximates $e^{-\lambda t}$ and is termed for this the *principal* root of the characteristic equation, while $z_2{}^j$ that bears no relationship to the true solution is termed the *spurious, parasitic, or extraneous* root of this equation.

Our main interest lies with systems of equations that become stiffer as $h \to 0$ and for them, the two step explicit method just presented is of no advantage over the simpler one step method of Euler. Indeed, the truncation error of the two step scheme in Eq. (8.112) is $O(\tau^3)$ as compared with the mere $O(\tau^2)$ in Euler's method, but stability imposes, in both cases, the same restriction $\tau < 2/\lambda_N$ on the time step size and with a large λ_N, it is stability, not accuracy, that dictates the step size.

The possibilities for creating other higher-order explicit multistep schemes by either including more steps or higher-order derivatives of y at the nodes are enormous, but owing to stability limitations, the practical usefulness of these schemes is small and we leave this matter now and turn our attention instead to *implicit* multistep methods.

We still strive to avoid the inclusion of $\ddot{y}$ in the time integration scheme because it makes the undesirable $KM^{-1}K$ appear in it, and set forth therefore to create a two step scheme that includes only the first time derivative of y, in the form

$$y_2 = \alpha_0 y_0 + \alpha_1 y_1 + \tau(\beta_0 \dot{y}_0 + \beta_1 \dot{y}_1 + \beta_2 \dot{y}_2) \qquad (8.122)$$

which is implicit if $\beta_2 \neq 0$ and explicit otherwise. The condition that Eq. (8.122) be exact for $y = 1$, $y = t$, $y = t^2$, and $y = t^3$ reduces Eq. (8.122) to

$$y_2 = \alpha y_0 + (1 - \alpha)y_1 + \tfrac{1}{12}\tau[\dot{y}_0(-1 + 5\alpha) + \dot{y}_1(8 + 8\alpha) + \dot{y}_2(5 - \alpha)] \qquad (8.123)$$

where α is a free parameter that is left from among α_0, α_1, β_0, β_1, and β_2. If, in addition, we ask Eq. (8.123) to be exact also for $y = t^4$, then $\alpha = 1$ and Eq. (8.123) becomes *Milne's* formula

$$y_2 = y_0 + \tfrac{1}{3}\tau(\dot{y}_0 + 4\dot{y}_1 + \dot{y}_2) \tag{8.124}$$

which for $Ky + \dot{y} = 0$ is

$$(I + \tfrac{1}{3}\tau K)y_2 = (I - \tfrac{1}{3}\tau K)y_0 - \tfrac{4}{3}\tau K y_1 \tag{8.125}$$

There is much to please us in Eq. (8.125): It is of high order—with truncation error $O(\tau^5)$—but only the simple matrix $I + \tfrac{1}{3}\tau K$ need be inverted in order to advance with it in time. So far so good, but unfortunately the scheme (8.125) is useless for our purpose; it is *unstable*. Indeed, for $\dot{y} + \lambda y = 0$, the characteristic equation for Milne's formula in Eq. (8.124) is

$$z^2(1 + \tfrac{1}{3}\psi) + \tfrac{4}{3}z\psi - 1 + \tfrac{1}{3}\psi = 0 \tag{8.126}$$

and near $\psi = 0$ its two roots are

$$z_1 = 1 - \psi, \qquad z_2 = -(1 + \tfrac{1}{3}\psi) \tag{8.127}$$

and $|z_2| > 1$ when $\psi > 0$.

To save the stability of this implicit two step scheme we sacrific accuracy, revoke the condition that it be exact for $y = t^4$, and use the free parameter α in Eq. (8.123) to make it stable. For this, we have to form the characteristic equation of the scheme in Eq. (8.123)

$$z^2[1 + \tfrac{1}{12}\psi(5 - \alpha)] + z[-1 + \alpha + \tfrac{2}{3}\psi(1 + \alpha)] + [-\alpha + \tfrac{1}{12}\psi(-1 + 5\alpha)] = 0 \tag{8.128}$$

When $\psi = 0$, its two roots are $z_1 = 1$ and $z_2 = -\alpha$ and a necessary condition for stability is that $|\alpha| < 1$ or $-1 < \alpha < 1$. The limits of stability are found from Eq. (8.128) to be

$$0 \leqslant \psi \leqslant 6(1 - \alpha)/(1 + \alpha) \tag{8.129}$$

and as $\alpha \to -1$, unconditional stability is approached. But at $\alpha = -1$, $z_2 = 1$ for all $\psi > 0$ and the scheme drops in accuracy. In fact, when $\alpha = -1$

$$z_1 = (1 - \tfrac{1}{2}\psi)/(1 + \tfrac{1}{2}\psi), \qquad z_2 = 1 \tag{8.130}$$

and

$$y_j = c_1 z_1{}^j + c_2 \tag{8.131}$$

When ψ is small and $y_1 = y_0 e^{-\psi}$, we find that $c_2 = \tfrac{1}{12}y_0\psi^2$ and as $j \to \infty$, $y_\infty \to O(\tau^2)$, not zero. For more on this scheme see Exercise 9.

A general theory has been developed by Dahlquist for the k-step scheme

$$y_{j+1} = \alpha_1 y_j + \cdots + \alpha_k y_{j-1+k} + \tau(\beta_0 \dot{y}_{j+1} + \cdots + \beta_k \dot{y}_{j-k+1}) \quad (8.132)$$

which is implicit if $\beta_0 \neq 0$ and is explicit otherwise. Its main results are stated in the following two theorems:

(i) *No explicit multistep method exists that is unconditionally stable.*
(ii) *No unconditionally stable multistep method has an error exceeding* $O(\tau^2)$.

Concerning theorem (i), it should be remarked that no other explicit scheme is known that is unconditionally stable, and the restriction on τ seems to be an inherent price we have to pay for explicitness. Theorem (ii) refers specifically to multistep methods that include only y and $\dot{y}$; the inclusion of $\ddot{y}$ or higher-order time derivatives of y lead, of course, to higher-order unconditionally stable schemes as in Eq. (8.79).

16. Predictor–Corrector Methods

The idea behind the predictor–corrector method is to use an *explicit* method to *predict* the next y, and then to use an *implicit* scheme to *correct* it, with the hope that accuracy and stability will be improved thereby. The variants of this method, and the possibilities for predictor–corrector interplay are enormous and we restrict ourselves to the representative case of Euler's predictor and Crank–Nickolson's corrector, such that

$$
\begin{aligned}
y_1' &= y_0 + \tau \dot{y}_0 &\text{(predict)} \\
y_1 &= y_0 + \tfrac{1}{2}\tau(\dot{y}_1' + \dot{y}_0) &\text{(correct)}
\end{aligned}
\qquad (8.133)
$$

We wish to analyze further the arrangement by which the predictor is used only once, but the corrector is successively applied in the cyclical form

$$y_1^{(k+1)} = y_0 + \tfrac{1}{2}\tau(\dot{y}_1^{(k)} + \dot{y}_0), \qquad (8.134)$$

where $y_1^{(0)} = y_0 + \tau \dot{y}_0$. To see what happens to the accuracy as $y_1^{(k)}$ is iterated by Eq. (8.134), we express $y_1^{(k+1)}$ backward in terms of y_0 and have that for $Ky + \dot{y} = 0$

$$y_1^{(k+1)} = (I - \tau K + \tfrac{1}{2}\tau^2 K^2 - \tfrac{1}{4}\tau^3 K^3 \pm \cdots) y_0 \qquad (8.135)$$

Comparing this with the Taylor expansion

$$y_1^{(k+1)} = (I - \tau K + \tfrac{1}{2}\tau^2 K^2 - \tfrac{1}{6}\tau^3 K^3 \pm \cdots) y_0 \qquad (8.136)$$

we see that accuracy is not gained after more than two cycles of Eq. (8.134), and hence from the accuracy viewpoint, two applications of the corrector is all we want. But we are anxious for stability and wish to see the effect of successive corrections on it. In particular, we entertain the hope that if the iterative scheme in Eq. (8.134) is carried out *to convergence*, the predictor–corrector procedure will become unconditionally stable as is the corrector. Unfortunately, this is not so.

To see why, we need the magnification factor for the scheme in Eq. (8.134), which for the one variable equation $\dot{y} + \lambda y = 0$ is obtained by adding $\frac{1}{2}\psi y_1^{(k+1)}$ to both sides of Eq. (8.134), resulting in

$$(1 + \tfrac{1}{2}\psi)y_1^{(k+1)} = y_0 - \tfrac{1}{2}\psi(y_1^{(k)} - y_1^{(k+1)} + y_0) \tag{8.137}$$

and then with $y_1^{(k)} - y_1^{(k+1)} = -2(-\tfrac{1}{2}\psi)^{k+2}y_0$ we obtain the desired relationship between $y_1^{(k)}$ and $y_{(0)}$ in the form

$$y_1^{(k)} = \left[\frac{1 - \tfrac{1}{2}\psi}{1 + \tfrac{1}{2}\psi} - 2\frac{(-\tfrac{1}{2}\psi)^{k+2}}{1 + \tfrac{1}{2}\psi}\right]y_0 \tag{8.138}$$

This magnification factor converges to that of the trapezoidal rule, *but only when $0 < \psi < 2$, which is the stability condition on Euler's predictor*. That the predictor–corrector method in Eq. (8.133), in its foregoing mode of application, is useless in extending the stability range of Euler's method is further demonstrated graphically in Fig. 8.11.

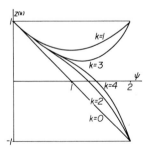

Fig. 8.11   Iterated magnification factor of the predictor–corrector scheme in Eq. (8.133).

Another way to look at the predictor–corrector method in Eq. (8.134) is to consider it as the iterative procedure

$$y_1^{(k+1)} = -\tfrac{1}{2}\tau K y_1^{(k)} + (I - \tfrac{1}{2}\tau K)y_0 \tag{8.139}$$

to solve

$$(I + \tfrac{1}{2}\tau K)y_1 = (I - \tfrac{1}{2}\tau K)y_0 \tag{8.140}$$

in Crank–Nickolson's scheme for y_1, with $-\tfrac{1}{2}\tau K$ being the iteration matrix. For its convergence, the spectral radius $\|-\tfrac{1}{2}\tau K\|_2$ of this matrix need be less than one, or $\lambda_N \tau < 2$, precisely the same restriction on τ that stability imposes in Euler's scheme.

17. Nonlinear Heat Conduction and the Runge–Kutta Method

A common source of nonlinearity in heat conduction is material properties that depend upon temperature. For instance, if the rod's conductivity grows exponentially with the temperature, then the equation describing the temperature distribution over it becomes

$$(e^y y')' = \alpha \dot{y} \tag{8.141}$$

which is nonlinear. The discretization of Eq. (8.141) by, say, finite differences is not difficult. In fact, for the jth node, we readily write for it the difference equation

$$h^{-2}\left[e^{y_a}y_{j-1} - (e^{y_a} + e^{y_b})y_j + e^{y_b}y_{j+1}\right] = \alpha \dot{y}_j$$

$$y_a = \tfrac{1}{2}(y_{j-1} + y_j), \qquad y_b = \tfrac{1}{2}(y_j + y_{j+1}) \tag{8.142}$$

and when all the interior equations have been written, and the boundary conditions introduced into them, we are left with the *nonlinear* system

$$K(y)y + \dot{y} = 0 \tag{8.143}$$

in which $K(y)$ is the temperature dependent global matrix.

Because of the nonlinearity of Eq. (8.143), implicit time integration schemes employed in its integration require the solution of a nonlinear algebraic system of equations for each step, not a very pleasant prospective. Explicit schemes, on the other hand, are as simple to use in the nonlinear case as in the linear case, except that the global stiffness matrix needs to be updated each step. Euler's scheme for Eq. (8.143) is

$$y_1 = \left[I - \tau K(y_0)\right]y_0 \tag{8.144}$$

and again no more than a matrix vector multiplication is required to pass from time level 0 to time level 1. Assessment of the stable time step τ is somewhat trickier in the nonlinear case since the maximal temperature has to be appraised beforehand in order to bound the maximum eigenvalue of $K(y)$. But if this estimate happens to be too liberal and τ too large, instability will manifest itself in the form of wild oscillations and a shorter time step should be taken.

Higher-order explicit one step schemes for Eq. (8.143) that include $\ddot{y}, \dddot{y} \ldots$ require the successive differentiation of $\dot{y} = -K(y)y$, which, because of the inclusion of y in K, might become unwieldy. This difficulty is ingeniously circumvented by the *Runge–Kutta* method that raises the order of the scheme—its accuracy by successive computations of $K(y)y$, instead of by successive differentiation. To present the method, we assume our nonlinear equation with only one variable of the form

$$\dot{y} = f(y), \qquad y(0) = y_0, \qquad t > 0 \tag{8.145}$$

where $f(y)$ is some function of y. Following Runge–Kutta, we write for it the *explicit* scheme

$$y_1 = y_0 + w_1 k_1 + w_2 k_2, \qquad k_1 = \tau f(y_0), \qquad k_2 = \tau f(y_0 + \alpha_{21} k_1) \tag{8.146}$$

in which w_1, w_2, and α_{21} are parameters to be determined for highest accuracy. To find them, we assume $f(y) = -\lambda y$ and have from Eq. (8.146) that

$$y_1 = [1 + \psi(w_1 + w_2) + w_2 \alpha_{21} \psi^2] y_0 \tag{8.147}$$

and upon comparison with Taylor's expansion, we decide that

$$w_1 + w_2 = 1, \qquad \alpha_{21} w_2 = \tfrac{1}{2} \tag{8.148}$$

There is some liberty in selecting w_1, w_2, and α_{21}, and if we choose $w_1 = w_2 = \tfrac{1}{2}$ and $\alpha_{21} = 1$, we get the second order Runge–Kutta scheme

$$y_1 = y_0 + \tfrac{1}{2}(k_1 + k_2), \qquad k_1 = \tau f(y_0), \qquad k_2 = \tau f(y_0 + k_1) \tag{8.149}$$

The possibilities for creating higher-order multistep explicit or implicit Runge–Kutta schemes are many, but we do not wish to continue and discuss them here in detail.

EXERCISES

1. Solve the heat conduction problem in Eq. (8.67), using the superstable scheme in Eq. (8.86). Choose $\tau = 0.05$ and compare your results with those in Fig. 8.6, obtained, using the trapezoidal rule.

2. Do the same as in Exercise 1 but use Calahan's scheme, given in Eq. (8.93), with $\alpha = -\sqrt{3}/3$.

3. Examine the stability and accuracy of the three step implicit scheme

$$y_3 = \alpha_0 y_0 + \alpha_1 y_1 + \alpha_2 y_2 + \tau(\beta_0 \dot{y}_0 + \beta_1 \dot{y}_1 + \beta_2 \dot{y}_2 + \beta_3 \dot{y}_3)$$

4. Use finite differences and Euler's method to solve the nonlinear heat conduction problem

$$(e^y y')' = \dot{y}, \qquad\qquad 0 < x < 1, \quad t > 0$$

$$y(x, 0) = 1, \qquad\qquad 0 < x < 1$$

$$y(0, t) = y(1, t) = 0, \qquad t > 0$$

5. Show that when the equation $\dot{y} = -\lambda(t)y, \lambda(t) > 0$, is solved, using the trapezoidal rule

$$y_1 = \frac{1 - \frac{1}{2}\tau\lambda(t_0)}{1 + \frac{1}{2}\tau\lambda(t_1)} y_0$$

then τ is restricted by stability consideration to

$$\tau[\lambda(t_0) - \lambda(t_1)] < 4$$

while the scheme

$$y_1 = \frac{1 - \frac{1}{2}\tau\lambda(t_{1/2})}{1 + \frac{1}{2}\tau\lambda(t_{1/2})} y_0$$

is unconditionally stable.
Study the accuracy of both schemes and extend your analysis to the nonlinear case $\dot{y} = -\lambda(y)y, \lambda(y) > 0$.

6. To avoid the temperature discontinuity at $x = 0$ in Eq. (8.67) modify the boundary condition $y(0, t) = 1$ to $y(0, t) = 1 - e^{-\beta t}$, $\beta > 0$, so that now

$$y'' = \dot{y}, \qquad 0 < x < 1, \quad t > 0$$

$$y(0, t) = 1 - e^{-\beta t}, \qquad y'(1, t) = 0$$

$$y(x, 0) = 0, \qquad 0 < x < 1$$

with exact solution

$$y(x, t) = 1 - e^{-\beta t} \frac{\cos(1 - x)\sqrt{\beta}}{\cos\sqrt{\beta}}$$

$$- \frac{16\beta}{\pi} \sum_{j=0}^{\infty} \frac{\exp[-(2j - 1)^2\pi^2 t/4]}{(2j - 1)[4\beta - \pi^2(2j - 1)^2]} \sin\left[\frac{(2j - 1)\pi x}{2}\right]$$

$$\beta \neq \left[\frac{\pi(2j - 1)}{2}\right]^2, \qquad j = 0, 1, 2, \ldots$$

Solve this exercise numerically as in Fig. 8.6 and observe the beneficial smoothing effect of $\beta > 0$. Start with $\beta = 2$.

7. Numerically solve the initial value problem $2\dot{y}y = 1$, $y(0) = 0$ using the scheme $y_1 = y_0 + \tau\dot{y}_1$ and compare your computed solution with the correct $y = \sqrt{t}$. Starting with the second step successively compute $\ddot{y}_1 = \tau^{-2}(y_0 - 2y_1 + y_2)$. Explain how $\ddot{y}_1$ can be used to control the step size in order to get a nearly constant truncation error at each step.

8. Solve $\dot{y} + y = 0$, $y(0) = 1$ with the scheme (8.104) starting with $y_0 = 1$ and $y_1 = z_1$ so as to have $c_1 = 1$, $c_2 = 0$ and $y_j = z_1{}^j$, $z_1 = -\psi + \sqrt{1 + \psi^2}$. Show that because of round-off errors c_2 is actually not zero and that the *numerical* solution is unstable.

9. Solve $\dot{y} + y = 0$, $y(0) = 1$ with the scheme (8.123) starting with $y_0 = 1$, $y_1 = e^{-\tau}$ and show that when $\alpha = -1$ and the scheme is unconditionally stable the computed $y(1)$ is only $O(\tau^2)$ in agreement with Dahlquist's theorem. What happens when $\alpha = -1 + O(\tau)$? Repeat your calculations with $\alpha = \frac{1}{2}$ and show that now that the scheme is only conditionally stable the accuracy of the scheme (8.123) is $O(\tau^3)$. Compare all this with the trapezoidal scheme.

SUGGESTED FURTHER READING

Carslaw, H. S., and Jaeger, J. C., *Conduction of Heat in Solids*. Oxford Univ. Press, London, 1948.

Creschino, F., and Kuntzman, J., *Numerical Solution of Initial Value Problems*. Prentice-Hall, Englewood Cliffs, New Jersey, 1966.

Richtmyer, R. D., and Morton, K. W., *Difference Methods for Initial-Value Problems*, 2nd ed. Wiley (Interscience), New York, 1967.

Lapidus, L., and Seinfeld, J. H., *Numerical Solution of Ordinary Differential Equations*. Academic Press, New York, 1971.

Gear, C. W., *Numerical Initial Value Problems in Ordinary Differential Equations*. Prentice-Hall, Englewood Cliffs, New Jersey, 1971.

Roberts, S. M., and Shipman, J. S., *Two-Point Boundary Value Problems: Shooting Methods*. Amer. Elsevier, New York, 1972.

Lambert, J. D., *Computational Methods in Ordinary Differential Equations*. Wiley, New York, 1973.

Willoughby, R. A., *Stiff Differential Systems*. Plenum, New York, 1974.

Shampine, L. D., and Gordon, M. K., *Computer Solution of Ordinary Differential Equations; the Initial Value Problem*. Freeman, San Francisco, California, 1975.

Widder, D. V., *The Heat Equation*. Academic Press, New York, 1975.

9 *Equation of Motion*

1. Spring–Mass System

The simple equation of motion of the spring and mass configuration in Fig. 9.1a will be referred to often in this chapter and we shall set it up. Consider again, for this purpose, Fig. 9.1a showing a spring of constant k, which is fixed to the left and with a mass m attached to its right end point. When the mass is displaced through the positive distance y to the right, the spring exerts on it a *restoring* force ky directed to the left, and Newton's second law of motion states here that

$$-ky = m\ddot{y} \tag{9.1}$$

or

$$\ddot{y} + \omega^2 y = 0, \qquad \omega^2 = k/m \tag{9.2}$$

With the two initial conditions of position and velocity

$$y(0) = y_0, \qquad \dot{y}(0) = \dot{y}_0 \tag{9.3}$$

Eqs. (9.2) and (9.3) constitute a linear second-order initial value problem, whose solution is

$$y = c_1 \sin \omega t + c_2 \cos \omega t \tag{9.4}$$

representing a bounded peroidic motion.

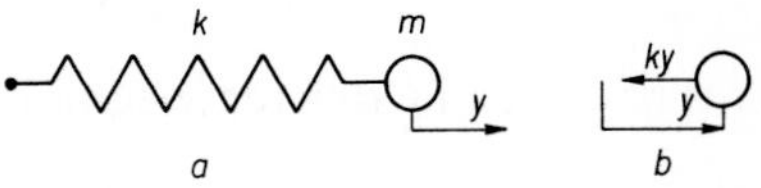

Fig. 9.1 A spring–mass system with one degree of freedom.

If the mass does not encounter friction during its motion and is acted upon by external forces derived from a potential, the sum of the potential and kinetic energies of the mass–spring system is *conserved* during the motion, or here

$$E = \tfrac{1}{2}ky^2 + \tfrac{1}{2}m\dot{y}^2 = \tfrac{1}{2}ky_0{}^2 + \tfrac{1}{2}m\dot{y}_0{}^2 \tag{9.5}$$

Indeed, Eq. (9.1) may be written as

$$m\frac{d\dot{y}}{dt}\frac{dy}{dy} + ky = 0 \tag{9.6}$$

Or with $\dot{y} = dy/dt$,

$$ky\,dy + m\dot{y}\,d\dot{y} = 0 \tag{9.7}$$

whose integration produces Eq. (9.5) with E as the constant of integration. Thus, the energy of the system is the *first integral* of the equation of motion and when such a first integral is available, it may be used in the numerical solution as a check on the accuracy of the computed solution.

2. Single Step Explicit Scheme

Probably the scheme that comes first to mind for the time integration of the second-order initial value problem is

$$\begin{aligned} y_1 &= y_0 + \tau\dot{y}_0 \\ \dot{y}_1 &= \dot{y}_0 + \tau\ddot{y}_0 \end{aligned} \tag{9.8}$$

which correctly predicts y_1 and $\dot{y}_1$ from the initial value y_0 and $\dot{y}_0$ when $y = 1$ and $y = t$, and in which τ is the time step. Let us examine the working of the scheme in Eq. (9.8) in the equation of motion $\ddot{y} + \omega^2 y = 0$. Here, because $\ddot{y} = -\omega^2 y$, the predictions of Eq. (9.8) become

$$\begin{aligned} y_1 &= y_0 + \tau\dot{y}_0 \\ \dot{y}_1 &= -\omega^2\tau y_0 + \dot{y}_0 \end{aligned} \tag{9.9}$$

or in general

$$\begin{aligned} y_{j+1} &= y_j + \tau\dot{y}_j \\ \dot{y}_{j+1} &= -\omega^2\tau y_j + \dot{y}_j \end{aligned} \tag{9.10}$$

This pair of finite difference equations is solved by

$$y_j = c_1 z^j, \qquad \dot{y}_j = c_2 z^j \tag{9.11}$$

where c_1 and c_2 are determined by the initial conditions. Substitution of y_j and $\dot{y}_j$ in Eq. (9.11) into the explicit one step scheme (9.10) produces the *pair* of characteristic equations

$$c_1 z = c_1 + c_2 \tau$$
$$c_2 z = -\omega^2 \tau c_1 + c_2 \tag{9.12}$$

which, upon the elimination of c_2 between them, become

$$z^2 - 2z + 1 + \psi = 0, \qquad \psi = \omega^2 \tau^2 \tag{9.13}$$

We expect $y_j = c_1 z^j$ to be oscillatory, the same way y in $\ddot{y} + \omega^2 y = 0$ is, which happens only when the magnification factor z is *complex*. In fact, the two roots of Eq. (9.13) are

$$z = 1 \pm i\psi^{1/2}, \qquad i^2 = -1 \tag{9.14}$$

and y_j is periodic. To bring this out more clearly we write the two roots z_1 and z_2 of Eq. (9.13) as

$$z_{1,2} = |z|(\cos\theta \pm i\sin\theta), \qquad |z| = (1 + \omega^2\tau^2)^{1/2} \tag{9.15}$$

and have from Eq. (9.14) that

$$\cos\theta = 1/|z|, \qquad \sin\theta = \omega\tau/|z| \tag{9.16}$$

Because there are *two* roots to the characteristic equation, y_j and $\dot{y}_j$ in Eq. (9.11) are written as

$$y_j = |z|^j(c_1 \cos j\theta + c_1{}' \sin j\theta)$$
$$\dot{y}_j = |z|^j(c_2 \cos j\theta + c_2{}' \sin j\theta) \tag{9.17}$$

To obtain the constants $c_1, c_1{}', c_2,$ and $c_2{}'$, we set $j = 0$ in both Eqs. (9.17) and readily have that $c_1 = y_0$ and $c_2 = \dot{y}_0$. Then we set $j = 1$ into these equations which become

$$y_1 = |z|(y_0 \cos\theta + c_1{}' \sin\theta)$$
$$\dot{y}_1 = |z|(\dot{y}_0 \cos\theta + c_2{}' \sin\theta) \tag{9.18}$$

and when y_1 and $\dot{y}_1$ in Eq. (9.18) are put into the difference Eqs. (9.9), they produce

$$|z|(y_0 \cos\theta + c_1{}' \sin\theta) = y_0 + \tau\dot{y}_0$$
$$|z|(\dot{y}_0 \cos\theta + c_2{}' \sin\theta) = -\omega^2\tau y_0 + \dot{y}_0 \tag{9.19}$$

But $|z| \cos\theta = 1$ and $|z| \sin\theta = \omega\tau$, and hence

$$c_1{}' = (1/\omega)\dot{y}_0, \qquad c_2{}' = -\omega y_0 \tag{9.20}$$

such that finally

$$y_j = (1 + \psi)^{j/2}(y_0 \cos j\theta + (1/\omega)\dot{y}_0 \sin j\theta),$$
$$\dot{y}_j = (1 + \psi)^{j/2}(\dot{y}_0 \cos j\theta - \omega y_0 \sin j\theta), \qquad \psi = \omega^2\tau^2 \qquad (9.21)$$

As for θ, we have from $\sin\theta = \omega\tau/|z|$ that when θ is small

$$\theta = \omega\tau(1 - \tfrac{1}{3}\psi), \qquad \psi = \omega^2\tau^2 \qquad (9.22)$$

and therefore

$$\theta j = \hat{\omega}t, \qquad \hat{\omega} = \omega(1 - \tfrac{1}{3}\psi) \qquad (9.23)$$

If we assume for the sake of brevity that $\dot{y}_0 = 0$, then

$$y_j = y_0(1 + \psi)^{j/2} \cos \hat{\omega}t, \qquad \psi = \omega^2\tau^2 \qquad (9.24)$$

as compared with the exact $y = y_0 \cos \omega t$.

The approximate solution in Eq. (9.24) is beset by frequency and amplitude errors. In the frequency, the error is seen by Eq. (9.23) to be $O(\omega^2\tau^2)$. To observe the error in the amplitude, we consider y_j and y at a fixed moment t, such that $j\tau = t$, and have that when $(j/\omega t)^2 \gg 1$,

$$|z|^j = e^{\omega^2 t\tau/2} \qquad (9.25)$$

As $\tau \to 0$, $|z|^j \to 1$, $\hat{\omega} \to \omega$, and *convergence* of the approximate solution y_j at time t to the exact $y(t)$ takes place. But for any τ, $|z|^j$ and with it the amplitude of the computed motion, grows exponentially with the time. Thus the scheme in Eq. (9.8) is *unstable* for any $\tau > 0$, and is consequently of little practical use. Physically interpreted, the difference scheme (9.8) includes an artificial energy source which drives the amplitude of y_j relentlessly up.

Another way to analyze the stability of the difference scheme (9.8) is to write it in the vector form $y_1 = Zy_0$, where $y_1{}^T = (y_1, \dot{y}_1)$ and

$$Z = \begin{bmatrix} 1 & \tau \\ -\omega\tau & 1 \end{bmatrix} \qquad (9.26)$$

In this notation $y_j = Z^j y_0$, and a necessary condition for stability is that the spectral radius $\rho(Z)$ of Z be less than 1. Here, $\rho(Z) = (1 + \omega^2\tau^2)^{1/2}$ and the scheme is unstable.

3.   Conditionally Stable Schemes

In the wake of our failure with the explicit single step scheme of the previous section, we propose to write another one, including this time $\ddot{y}$, in the form

$$y_1 = y_0 + \tau \dot{y}_0 + \tfrac{1}{2}\tau^2 \ddot{y}_0$$
$$\dot{y}_1 = \dot{y}_0 + \tau[\alpha \ddot{y}_0 + (1 - \alpha)\ddot{y}_1]$$

(9.27)

in which α is a parameter that we shall use to stabilize the scheme. For the linear $\ddot{y} + \omega^2 y = 0$, this scheme is explicit even if $\alpha \neq 1$ since $\ddot{y}_1 = -\omega^2 y_1$, and y_1 is already computed by the first of Eqs. (9.27). However, if the equation of motion is nonlinear in the velocity, the scheme (9.27) becomes *implicit*.

To determine its stability, we introduce $\ddot{y} = -\omega^2 y$ into the scheme (9.27) and get with $\psi = \omega^2 \tau^2$ that

$$y_1 = y_0(1 - \tfrac{1}{2}\psi) + \tau \dot{y}_0$$
$$\dot{y}_1 = -y_0(1/\tau)\psi[1 + \tfrac{1}{2}\psi(\alpha - 1)] + \dot{y}_0[1 - (1 - \alpha)\psi]$$

(9.28)

which, again, admits a solution of the form $y_j = c_1 z^j$ and $\dot{y}_j = c_2 z^j$, leading, as before, to the pair of characteristic equations

$$c_1 z = c_1(1 - \tfrac{1}{2}\psi) + \tau c_2$$
$$c_2 z = -c_1(1/\tau)\psi[1 + \tfrac{1}{2}\psi(\alpha - 1)] + c_2[1 - (1 - \alpha)\psi]$$

(9.29)

and then to

$$z^2 - 2z[1 - \tfrac{1}{2}\psi(\tfrac{3}{2} - \alpha)] + 1 + \psi(\alpha - \tfrac{1}{2}) = 0$$

(9.30)

For z to be complex, the discriminant of Eq. (9.30) must be negative, or

$$0 < \tau < \frac{1}{\omega} \frac{2}{|\tfrac{3}{2} - \alpha|}$$

(9.31)

Moreover, according to Descartes' rule, the free term in a quadratic equation equals the product of its two roots, and hence here

$$|z|^2 = 1 + \psi(\alpha - \tfrac{1}{2})$$

(9.32)

After j steps, the amplitude of the computed motion is proportional to $|z|^j$ or, according to Eq. (9.32), to $[1 + \psi(\alpha - \tfrac{1}{2})]^{j/2}$. When $\alpha > \tfrac{1}{2}$, the amplitude of the computed motion *increases* with j, the scheme becomes energy *generating* and unstable. When $\alpha < \tfrac{1}{2}$, $|z| < 1$ and the amplitude of the computed motion *diminishes* as time elapses. In this case, the scheme includes a hidden *energy sink*, or an *artificial viscosity* which consumes energy and drives the amplitude down. Only when $\alpha = \tfrac{1}{2}$ is the scheme *energy conserving* so that the amplitude of the computed motion remains constant. Figure 9.2 shows the computed y for $y + \ddot{y} = 0$, $y(0) = 1$, and $\dot{y}(0) = 0$, followed in

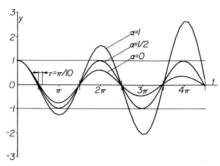

Fig. 9.2   Solution of $\ddot{y} + y = 0$, $y(0) = 1$, $\dot{y}(0) = 0$, with the scheme in Eq. (9.27) for different values of α.

time with the scheme (9.27) using the different values $\alpha = 0$, $\alpha = \frac{1}{2}$, $\alpha = 1$, and $\tau = \pi/10$.

When $\alpha = \frac{1}{2}$, the scheme in Eq. (9.27) becomes

$$y_1 = y_0 + \tau\dot{y}_0 + \tfrac{1}{2}\tau^2\ddot{y}_0$$
$$\dot{y}_1 = \dot{y}_0 + \tfrac{1}{2}\tau(\ddot{y}_0 + \ddot{y}_1)$$

(9.33)

with τ limited by stability to

$$0 < \tau < 2/\omega$$

(9.34)

The complex conjugate roots of the characteristic equation (9.30) are now

$$z_{1,2} = 1 - \tfrac{1}{2}\psi \pm i\omega\tau(1 - \tfrac{1}{4}\omega^2\tau^2)^{1/2}$$

(9.35)

such that if $\dot{y}_0 = 0$, then

$$y_j = y_0 \cos j\theta, \qquad \dot{y}_j = -y_0\omega(1 - \tfrac{1}{4}\psi)^{1/2}\sin j\theta$$

(9.36)

where

$$\sin\theta = \omega\tau(1 - \tfrac{1}{4}\psi)^{1/2}$$

(9.37)

As $\tau \to 0$ so does $\sin\theta$, and for a small $\psi = \omega^2\tau^2$

$$\theta = \omega\tau(1 + \tfrac{1}{24}\omega^2\tau^2)$$

(9.38)

implying that the error in the frequency with the scheme (9.33) is $O(\tau^2\omega^2)$.

Because it includes only the velocity $\dot{y}$, the acceleration $\ddot{y}$, and no higher time derivatives of y, the scheme in Eq. (9.33) is computationally attractive and is indeed one of the most widely used explicit time integration schemes to solve the equation of motion. However, it becomes *implicit* when the equation of motion includes a nonlinear term in the velocity.

In our search for a scheme that is always explicit, we try

$$y_1 = y_0 + \tau\dot{y}_0 + \tfrac{1}{2}\tau^2\ddot{y}_0 + \alpha\tau^3\dddot{y}_0 + \beta\tau^4\ddddot{y}_0$$
$$\dot{y}_1 = \dot{y}_0 + \tau\ddot{y}_0 + \tfrac{1}{2}\tau^2\dddot{y}_0 + \beta'\tau^3\ddddot{y}_0$$

(9.39)

which for $\ddot{y} + \omega^2 y = 0$ becomes

$$y_1 = y_0(1 - \tfrac{1}{2}\psi + \beta\psi^2) + \tau\dot{y}_0(1 - \alpha\psi),$$
$$\dot{y}_1 = (1/\tau)y_0(\beta'\psi^2 - \psi) + \dot{y}_0(1 - \tfrac{1}{2}\psi), \qquad \psi = \omega^2\tau^2 \qquad (9.40)$$

with which the characteristic equation

$$z^2 - 2z(1 - \tfrac{1}{2}\psi + \tfrac{1}{2}\beta\psi^2) + 1 + \psi^2(\beta - \beta' - \alpha + \tfrac{1}{4}) + \psi^3(\alpha\beta' - \tfrac{1}{2}\beta) = 0 \qquad (9.41)$$

is associated. For the scheme (9.39) to be conserving and the solution periodic, the roots of Eq. (9.41) have to be complex and of modulus one. By Descartes' rule, the free term of Eq. (9.41) must be one for this to happen independently of ψ, and we are restricted to

$$\beta - \beta' - \alpha + \tfrac{1}{4} = 0, \qquad \alpha\beta' - \tfrac{1}{2}\beta = 0 \qquad (9.42)$$

Elimination of α between these equations leaves us with

$$\beta'^2 - \beta'(\tfrac{1}{4} + \beta) + \tfrac{1}{2}\beta = 0 \qquad (9.43)$$

and for a real β', β is restricted to $\beta \leq (3 - \sqrt{8})/4$ and $\beta \geq (3 + \sqrt{8})/4$. To choose β, we perform an error analysis on the frequency and find that

$$\hat{\omega} = \omega[1 + \psi(\tfrac{1}{24} - \tfrac{1}{2}\beta)] \qquad (9.44)$$

We cannot reduce the error in ω to lower than $O(\tau^2)$ since for this we have to choose $\beta = \tfrac{1}{12}$, which is forbidden. To minimize the coefficient of ψ in Eq. (9.44), we choose $\beta = \tfrac{1}{24}$, which is only slightly less than $(3 - \sqrt{8})/4$. But we could equally well choose $\beta = 0$, and consequently $\beta' = 0$ and $\alpha = \tfrac{1}{4}$, which is slightly less accurate, but which considerably simplifies the scheme (9.39), that is now

$$y_1 = y_0 + \tau\dot{y}_0 + \tfrac{1}{2}\tau^2\ddot{y}_0 + \tfrac{1}{4}\tau^3\dddot{y}_0$$
$$\dot{y}_1 = \dot{y}_0 + \tau\ddot{y}_0 + \tfrac{1}{2}\tau^2\dddot{y}_0 \qquad (9.45)$$

Figure 9.3 shows the error in the solution of $\ddot{y} + y = 0$, computed with the explicit scheme (9.39) and with different α, β, and β' values.

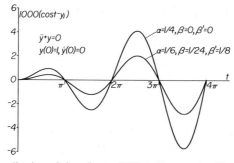

Fig. 9.3 Application of the scheme (9.39) to $\ddot{y} + y = 0$, $y(0) = 1$, $\dot{y}(0) = 0$.

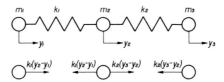

Fig. 9.4   One-dimensional array of springs and masses.

4.   Lattice of Springs and Masses

Figure 9.4 shows a one-dimensional *lattice* of masses attached by springs, which could be a simple model for the arrangement of atoms in a crystal, an elastic rod, or an air column in a tube, the motion of which we want to study. To form the equation of motion of this system, we isolate the masses, consider the forces acting on them, and write Newton's second law of motion for each one of them. Regarding Fig. 9.4, we write these equations as

$$m_1 \ddot{y} = k_1(y_2 - y_1)$$

$$m_2 \ddot{y}_2 = k_2(y_3 - y_2) - k_1(y_2 - y_1) \qquad (9.46)$$

$$m_3 \ddot{y}_3 = -k_2(y_3 - y_2)$$

or in matrix form

$$Ky + M\ddot{y} = 0 \qquad (9.47)$$

where $y^T = (y_1, y_2, y_3)$, and where

$$K = \begin{bmatrix} k_1 & -k_1 & \\ -k_1 & k_1 + k_2 & -k_2 \\ & -k_2 & k_2 \end{bmatrix}, \qquad M = \begin{bmatrix} m_1 & & \\ & m_2 & \\ & & m_3 \end{bmatrix} \qquad (9.48)$$

are the global stiffness and mass matrices of the elastic system in Fig. 9.4.

The elastic and kinetic energies stored in the mass–spring lattice are

$$E(y, y) = \tfrac{1}{2}[k_1(y_2 - y_1)^2 + k_2(y_3 - y_2)^2] \qquad (9.49)$$

and

$$K(\dot{y}, \dot{y}) = \tfrac{1}{2}[m_1 \dot{y}_1^2 + m_2 \dot{y}_2^2 + m_3 \dot{y}_3^2] \qquad (9.50)$$

respectively, such that

$$E(y, y) = \tfrac{1}{2}y^T K y, \qquad K(\dot{y}, \dot{y}) = \tfrac{1}{2}\dot{y}^T M \dot{y} \qquad (9.51)$$

From Eq. (9.46), we readily conclude that in this case K is positive *semidefinite* since $y_1 = y_2 = y_3$ is a rigid body mode that stores no energy in the springs.

Had the lattice in Fig. 9.4 been anchored at one end, the rigid body mode would have been excluded, resulting in a positive definite global stiffness matrix K.

By writing $x = M^{1/2}y$, where $M^{1/2}M^{1/2} = M$, Eq. (9.47) is brought to the more convenient form

$$K'x + \ddot{x} = 0, \qquad K' = M^{-1/2}KM^{-1/2} \tag{9.52}$$

to which any one of the previous time integration schemes may be applied. Concerning the scheme (9.45), it is interesting that even though it includes $\ddot{y}$, it does not entail the appearance of matrix products when applied to the system of second-order equations (9.52). In fact, in this case

$$x_1 = (I - \tfrac{1}{2}\tau^2 K')x_0 + \tau(I - \tfrac{1}{4}\tau^2 K')\dot{x}_0$$
$$\dot{x}_1 = -\tau K'x_0 + (I - \tfrac{1}{2}\tau^2 K')\dot{x}_0 \tag{9.53}$$

which involves at each step only vector matrix multiplications. But if K' is a constant matrix, as it is in all linear problems, then the scheme in Eq. (9.33) which assumes here the form

$$x_1 = (I - \tfrac{1}{2}\tau^2 K')x_0 + \tau\dot{x}_0$$
$$\dot{x}_1 = \dot{x}_0 - \tfrac{1}{2}\tau K'(x_0 + x_1) \tag{9.54}$$

is more efficient.

5. Modal Decomposition

Modal or Fourier analysis can be applied to the second-order system $Ky + M\ddot{y} = 0$ in the same way it was applied in Chapter 8 to the first-order system arising in heat conduction. Once more we expand the *vector* $y(x)$ in terms of the eigenvectors of $Kx = \lambda Mx$ in the form $y(t) = c_1 x_1 + c_2 x_2 + \cdots + c_N x_N$, and get upon substitution into $Ky + M\ddot{y} = 0$ that

$$\sum_{j=1}^{N} (\ddot{c}_j + \lambda_j c_j)Mx_j = 0 \tag{9.55}$$

Inasmuch as Mx_j are linearly independent,

$$\ddot{c}_j + \lambda_j c_j = 0, \qquad j = 1, 2, \ldots, N \tag{9.56}$$

and

$$c_j = a_j \cos \omega_j t + b_j \sin \omega_j t \tag{9.57}$$

where $\omega_j{}^2 = \lambda_j$ and where the constants a_j and b_j are obtained from the initial conditions of position and velocity.

The eigenvectors $x_1, x_2, \ldots, x_N$ of $Kx = \omega^2 Mx$ are the *natural modes of vibration* of the system and the corresponding $\omega_1, \omega_2, \ldots, \omega_N$ are the natural frequencies of the same system.

As an example of modal decomposition, let us take the spring–mass system of Fig. 9.4 with $k_1 = k_2 = 1$ and $m_1 = m_2 = m_3 = 1$, such that $M = I$ and K is

$$K = \begin{bmatrix} 1 & -1 & \\ -1 & 2 & -1 \\ & -1 & 1 \end{bmatrix} \tag{9.58}$$

We readily compute for this the three eigenvectors $x_1{}^T = (1,1,1)$, $x_2{}^T = (1, 0, -1)$, and $x_3{}^T = (1, -2, 1)$ shown in Fig. 9.5. Corresponding to these, we have $\lambda_1 = 0, \lambda_2 = 1$, and $\lambda_3 = 3$. Now

$$\begin{bmatrix} y_1(t) \\ y_2(t) \\ y_3(t) \end{bmatrix} = (a_1 + b_1 t) \begin{bmatrix} 1 \\ 1 \\ 1 \end{bmatrix} + (a_2 \cos t + b_2 \sin t) \begin{bmatrix} 1 \\ 0 \\ -1 \end{bmatrix}$$

$$+ (a_3 \cos \sqrt{3} t + b_3 \sin \sqrt{3} t) \begin{bmatrix} 1 \\ -2 \\ 1 \end{bmatrix} \tag{9.59}$$

and

$$\begin{bmatrix} \dot{y}_1(t) \\ \dot{y}_2(t) \\ \dot{y}_3(t) \end{bmatrix} = b_1 \begin{bmatrix} 1 \\ 1 \\ 1 \end{bmatrix} + (-a_2 \sin t + b_2 \cos t) \begin{bmatrix} 1 \\ 0 \\ -1 \end{bmatrix}$$

$$+ (-a_3 \sqrt{3} \sin \sqrt{3} t + b_3 \sqrt{3} \cos \sqrt{3} t) \begin{bmatrix} 1 \\ -2 \\ 1 \end{bmatrix} \tag{9.60}$$

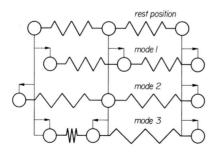

Fig. 9.5   Three natural (normal) modes of vibration of the lattice in Fig. 9.4.

Assuming the initial conditions to be $y_0^T = (1, 3, -2)$ and $\dot{y}_0^T = (0, 0, 0)$, we have from Eqs. (9.59) and (9.60) that

$$
\begin{bmatrix} 1 \\ 3 \\ -2 \end{bmatrix} = a_1 \begin{bmatrix} 1 \\ 1 \\ 1 \end{bmatrix} + a_2 \begin{bmatrix} 1 \\ 0 \\ -1 \end{bmatrix} + a_3 \begin{bmatrix} 1 \\ -2 \\ 1 \end{bmatrix} \tag{9.61}
$$

and

$$
\begin{bmatrix} 0 \\ 0 \\ 0 \end{bmatrix} = b_1 \begin{bmatrix} 1 \\ 1 \\ 1 \end{bmatrix} + b_2 \begin{bmatrix} 1 \\ 0 \\ -1 \end{bmatrix} + \sqrt{3} b_3 \begin{bmatrix} 1 \\ -2 \\ 1 \end{bmatrix} \tag{9.62}
$$

To fix a_1, a_2, a_3, b_2, and b_3, we multiply Eqs. (9.61) and (9.62) by the eigenvectors x_1^T, x_2^T, and x_3^T and have that $a_1 = \frac{2}{3}$, $a_2 = \frac{3}{2}$, $a_3 = -\frac{7}{6}$, $b_1 = 0$, $b_2 = 0$, and $b_3 = 0$, such that finally

$$
\begin{bmatrix} y_1(t) \\ y_2(t) \\ y_3(t) \end{bmatrix} = \frac{2}{3} \begin{bmatrix} 1 \\ 1 \\ 1 \end{bmatrix} + \frac{3}{2} \cos t \begin{bmatrix} 1 \\ 0 \\ -1 \end{bmatrix} - \frac{7}{6} \cos \sqrt{3} t \begin{bmatrix} 1 \\ -2 \\ 1 \end{bmatrix} \tag{9.63}
$$

Modal analysis is attractive when all eigenvalues and eigenvectors of $Ky = \lambda My$ are inexpensive obtainable. Otherwise, a step-by-step procedure for the integration in time is preferred.

6.   Stability Conditions for $Ky + M\ddot{y} = 0$

When applied to the system of equations $Ky + M\ddot{y} = 0$, the explicit single step scheme (9.33) assumes the form

$$
\begin{aligned}
My_1 &= (M - \tfrac{1}{2}\tau^2 K)y_0 + \tau M \dot{y}_0 \\
M\dot{y}_1 &= -\tfrac{1}{2}\tau K(y_0 + y_1) + M\dot{y}_0
\end{aligned} \tag{9.64}
$$

in which $y_0, y_1, \dot{y}_0$, and $\dot{y}_1$ are vectors of length N. The solution to Eq. (9.64) may be written in the form

$$
\begin{aligned}
y_j &= c_1 z_1{}^j x_1 + c_2 z_2{}^j x_2 + \cdots + c_N z_N{}^j x_N \\
\dot{y}_j &= c_1' z_1{}^j x_1 + c_2' z_2{}^j x_2 + \cdots + c_N' z_N{}^j x_N
\end{aligned} \tag{9.65}
$$

where $x_1, x_2, \ldots, x_N$ are the eigenvectors of $Kx = \omega^2 Mx$, and where $c_1, c_2, \ldots, c_N$ and $c_1', c_2', \ldots, c_N'$ are constants determined by the initial conditions. Substitution of Eq. (9.65) into the difference equation (9.64) yields

$$\sum_{j=1}^{N} Mx_j(c_j z_j - c_j + \tfrac{1}{2}\lambda_j \tau^2 c_j - \tau c_j') = 0$$

$$\sum_{j=1}^{N} Mx_j(c_j' z_j + \tfrac{1}{2}\tau\lambda_j c_j + \tfrac{1}{2} z_j \tau \lambda_j c_j - c_j') = 0 \qquad (9.66)$$

Since all the eigenvectors $x_1, x_2, \ldots, x_N$ are linearly independent and M is positive definite, all terms in the parentheses in Eqs. (9.66) must vanish and we get the system of characteristic equations

$$c_j(z_j - 1 + \tfrac{1}{2}\psi_j) - \tau c_j' = 0,$$
$$\tfrac{1}{2}c_j \psi_j(1 + z_j) + c_j' \tau(z_j - 1) = 0, \qquad \psi_j = (\omega_j \tau)^2 \qquad (9.67)$$

corresponding to (9.29) when there is only one degree of freedom in the oscillating system. Elimination of c_j and c_j' between Eqs. (9.67) leaves us the system of characteristic equations

$$z_j^2 - 2z_j(1 - \tfrac{1}{2}\psi_j) + 1 = 0, \qquad j = 1, 2, \ldots, N \qquad (9.68)$$

and for all computed modes to oscillate harmonically, z_j must be complex or $|1 - \tfrac{1}{2}\psi_j| < 1$, which means that

$$0 < \tau < 2/\max_j \{\omega_j\} \qquad (9.69)$$

In a stiff system, stability can impose a severe restriction on the maximum size of τ. Consider, for instance, a spring–mass system used to produce a discrete model of an elastic rod. In this, the rod is divided into equal segments of size h with the mass m of each segment concentrated at its center and the elastic link between them represented by a spring of constant k. The segment mass m is then proportional to h, and we assume $m = h$, while the stiffness k of the linking spring is proportional to $1/h$, and we assume simply $k = 1/h$. With this mass and stiffness distribution, the global mass matrix becomes $M = hI$ and the global stiffness matrix is now

$$K = \frac{1}{h}
\begin{bmatrix}
1 & -1 & & & & & \\
-1 & 2 & -1 & & & & \\
 & -1 & 2 & -1 & & & \\
 & & \cdot & \cdot & \cdot & & \\
 & & & -1 & 2 & -1 & \\
 & & & & -1 & 2 & -1 \\
 & & & & & -1 & 1
\end{bmatrix} \qquad (9.70)$$

Gerschgorin's and Rayleigh's theorems predict that for this $Ky = \lambda hy$

$$2/h^2 \leqslant \lambda_N \leqslant 4/h^2 \tag{9.71}$$

and consequently, according to Eq. (9.69), the stable τ is limited here to $0 < \tau < h$.

In stiff systems, stability is determined by the *highest* natural frequency $\omega_N{}^2 = \lambda_N$ present in the elastic system, even though this mode may contribute only insignificantly to the description of its motion. How this happens is discussed in the next detailed example.

Consider a lattice as in Fig. 9.4, but fixed at one end, consisting of 25 masses, all of mass $m = h$, joined by 25 springs, all of stiffness $k = 1/h$, where $h = \frac{1}{25}$. The system of equations of motion describing the movement of this elastic system is $Ky + \ddot{y} = 0$, where

$$K = \frac{1}{h^2} \begin{bmatrix} 2 & -1 & & & & \\ -1 & 2 & -1 & & & \\ & & \cdot & \cdot & \cdot & \\ & & & \cdot & \cdot & \cdot \\ & & & & \cdot & \cdot & \cdot \\ & & & & -1 & 2 & -1 \\ & & & & & -1 & 1 \end{bmatrix} \tag{9.72}$$

We know the natural frequencies of this lattice to be

$$\omega_j = (2/h)\sin\tfrac{1}{4}\pi h(2j - 1) \tag{9.73}$$

or when $\pi h/4 \ll 1$,

$$\pi/2 \leqslant \omega_j \leqslant 2/h \tag{9.74}$$

In our present example, $h = \frac{1}{25}$ and hence $\pi/2 \leqslant \omega_j \leqslant 50$, and for stability, if we intend to use the scheme (9.33), we have to restrict ourselves to $\tau \leqslant 0.04$.

The eigenvector x_1, corresponding to the fundamental natural frequency ω_1, is known to be

$$x_1{}^T = (\sin\theta, \sin 2\theta, \dots, 1), \qquad \theta = \pi h/2 \tag{9.75}$$

Suppose that the initial set of displacements imparted to the lattice is precisely $x_1, y(0) = y_0 = x_1$, and that the masses are released with zero initial velocity. In this case, only x_1 is involved in describing the subsequent motion of the system, and

$$y(t) = x_1 \cos\omega_1 t, \qquad \omega_1 \cong \pi/2 \tag{9.76}$$

meaning that all the masses oscillate with the *same* frequency $\omega_1 \cong \pi/2$ and that the amplitudes of oscillation of the different masses are proportional to the components of the fundamental eigenvector x_1. Theoretically, the

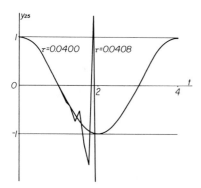

Fig. 9.6   Motion of the 25th mass in a spring–mass array fixed at one point and given an initial displacement in the form of its first natural mode of vibration. Advance in time is carried out with scheme (9.33).

computed solution is, according to Eq. (9.65), only

$$y_j = c_1 z_1{}^j x_1 \qquad\qquad (9.77)$$

and only z_1 need be complex and of modulus one for the computed motion to be conserving and oscillatory, or $\tau < 2/\omega_1 = 4$. In reality it is all otherwise. Because of the round-off errors in the computer, the *numerical* value of y_0 is slightly different from x_1, and consequently the eigenvector expansion of the perturbed y_0 includes not only x_1, but actually all the rest of the eigenvectors of K. As a result, y_j is not as in Eq. (9.77), but rather as in Eq. (9.65). Admittedly, the coefficients $c_2, c_3, \ldots, c_N$ in this expansion are all much smaller than c_1, but if z_j is real for some j, $|z_j| > 1$, and $|z_j|^j$ grows until it eventually destroys the computed solution. The swiftness with which this instability can occur is shown in Fig. 9.6, referring to the solution of $Ky + \ddot{y} = 0$ with K in Eq. (9.70) and with the scheme (9.33), starting with $y_0 = x_1$. Computation was carried out twice, first with $\tau = 0.04$, which is nearly the stability limit on the size of τ, and then with $\tau = 0.0408$, which is slightly above it. For the latter choice of τ, $z_{25} = -1.5$ and the influence of $c_{25}(-1.5)^j x_N$ becomes destructive at about a quarter of the fundamental period.

7.   Nonlinear Equation of Motion

The stability analysis of the previous section is confined to the *linear* equation of motion. No such general elementary analysis is available for the nonlinear equation of motion. But if the nonlinearity is slight, we expect its stability conditions to be nearly those for the linearized equation. A choice of a too large τ, causing instability, does not take long to detect and correct. Also, if the physical system is conserving energy, or momentum, or any other first integral is known, this may serve as a computational check on the conservation of the time integration scheme.

We shall consider now the numerical solution of some nonlinear equations of motion. More are to be found in the Exercises. It is customary in structural mechanics to classify the nonlinearities into (i) *material* and (ii) *geometrical*, where the first arises from spring materials that do not obey *Hook's* law and for which the stretching is not proportional to the applied force, while the second comes from large geometrical displacements but with possibly linear elastic materials. A simple example of material nonlinearity is provided by a spring of constant

$$k = k_0(1 + \varepsilon y^2) \tag{9.78}$$

such that the force–displacement relation is now

$$f(y) = k_0 y(1 + \varepsilon y^2) \tag{9.79}$$

When $\varepsilon > 0$, the spring becomes *stiffer* with the displacement, while when $\varepsilon < 0$, it becomes *softer* (more flexible) with stretching. With a restoring force as in Eq. (9.79), the equation of motion of a one mass–spring system becomes

$$m\ddot{y} + k_0 y(1 + \varepsilon y^2) = 0 \tag{9.80}$$

and a first integral E is readily obtained for it in the form

$$E = \tfrac{1}{2}m\dot{y}^2 + \tfrac{1}{2}k_0 y^2(1 + \tfrac{1}{2}\varepsilon y^2) \tag{9.81}$$

We propose to numerically solve Eq. (9.80) by the scheme

$$
\begin{aligned}
y_1 &= y_0 + \tau\dot{y}_0 - \tfrac{1}{2}\tau^2 f(y_0) \\
\dot{y}_1 &= \dot{y}_0 - \tfrac{1}{2}\tau[f(y_0) + f(y_1)]
\end{aligned} \tag{9.82}
$$

adapted from Eq. (9.33), which has been shown to be conserving in the linear case. Figure 9.7 shows the results of this computation, done with $\varepsilon = 1$, $k_0 = 1$, $m = 1$, $y_0 = 1$, $\dot{y}_0 = 0$, and $\tau = \pi/20$, including the energy error that suggests that here also the numerical scheme is energy conserving.

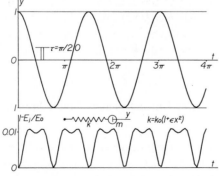

Fig. 9.7 Mass–spring system with a nonlinear spring. E_j denotes the computed total energy of the system at the jth time step.

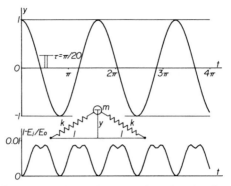

Fig. 9.8 Mass–spring system with nonlinear equation of motion due to large displacement of *m*.

A simple example of geometric nonlinearity comes from the mass-between-two-springs system shown in Fig. 9.8. The springs are assumed to be under an *initial* tension p and to be linear, such that when the mass is *laterally* displaced by the amount y, the springs stretch by the amount

$$s = (l^2 + y^2)^{1/2} - l \tag{9.83}$$

causing a restoring force $f(y)$ in the y direction

$$f(y) = 2p \frac{y}{s+l} + 2ky\left(1 - \frac{l}{s+l}\right) \tag{9.84}$$

The sum E of the elastic energy stored in the *two* springs and the kinetic energy of the mass is

$$E = 2ps^2 + ks^2 + \tfrac{1}{2}m\dot{y}^2 \tag{9.85}$$

where s is related to y in Eq. (9.83).

The numerical solution of the equation of motion of this system is carried out with the explicit scheme (9.82) for $l = \tfrac{1}{2}$, $k = 1$, $m = 1$, $p = \tfrac{1}{4}$, $y_0 = 1$, $\dot{y}_0 = 0$, and $\tau = \pi/20$, and the results of this computation are shown in Fig. 9.8, which is very similar to Fig. 9.7.

8.   Single Step Unconditionally Stable Implicit Scheme

Explicit schemes are computationally attractive for linear systems of second-order equations because no matrix inversion (except for M^{-1}, which is avoided by lumping) is required in them. There is, however, a price to pay for this: *conditional* stability, which may be a steep one in the presence of frequencies that are high relative to the fundamental. For in this case, the step size τ is severely limited by stability considerations, forcing us to march in time with steps that may well be considerably below those needed for the desired accuracy.

In this section we shall set up an unconditionally stable time integration scheme for the second-order equation of motion, but to achieve unconditional stability we shall have to accept *implicitness* in the scheme. Initially, the scheme is written in the form

$$y_1 = y_0 + \tau \dot{y}_0 + \tfrac{1}{2}\tau^2[\alpha \ddot{y}_0 + (1 - \alpha)\ddot{y}_1]$$
$$\dot{y}_1 = \dot{y}_0 + \tau[(1 - \beta)\ddot{y}_0 + \beta \ddot{y}_1]$$
(9.86)

in which α and β are determined by conservation of energy and unconditional stability conditions. For $\ddot{y} + \omega^2 y = 0$, the scheme in Eq. (9.86) gives rise to the characteristic equation

$$z^2 - z\,\frac{2 + \psi(\tfrac{1}{2} - \alpha - \beta)}{1 + \tfrac{1}{2}\psi(1 - \alpha)} + \frac{1 + \psi(1 - \beta - \tfrac{1}{2}\alpha)}{1 + \tfrac{1}{2}\psi(1 - \alpha)} = 0$$
(9.87)

and we seek to adjust α and β to make the roots of this equation complex conjugates of modulus one. To make $|z| = 1$ for any ψ, the last term in Eq. (9.87) must be 1, and this occurs with $\beta = \tfrac{1}{2}$. For z to be complex, we must have

$$\left|\frac{1 - \tfrac{1}{2}\alpha\psi}{1 + \tfrac{1}{2}\psi(1 - \alpha)}\right| < 1$$
(9.88)

which occurs with $\alpha = \tfrac{1}{2}$, and we have the scheme

$$y_1 = y_0 + \tau \dot{y}_0 + \tfrac{1}{4}\tau^2(\ddot{y}_0 + \ddot{y}_1)$$
$$\dot{y}_1 = \dot{y}_0 + \tfrac{1}{2}\tau(\ddot{y}_0 + \ddot{y}_1)$$
(9.89)

which consists of only a single step, and is implicit, unconditionally stable, and conserving. In the engineering literature it is often named after *Newmark*.

To assess the frequency error that results from Newmark's method, we observe that with $\alpha = \beta = \tfrac{1}{2}$, the roots of the characteristic equation (9.87) are such that if $z = \cos\theta \pm i\sin\theta$, then

$$\sin\theta = \psi^{1/2}/(1 + \tfrac{1}{4}\psi), \qquad \psi = \omega^2\tau^2$$
(9.90)

When θ is small, we may write

$$\theta = \omega\tau(1 - \tfrac{1}{12}\psi)$$
(9.91)

and the frequency error in Newmark's method is $O(\omega^2\tau^2)$.

Applied to the second-order system $Ky + \ddot{y} = 0$, the scheme in Eq. (9.89) becomes

$$(I + \tfrac{1}{4}\tau^2 K)y_1 = (I - \tfrac{1}{4}\tau^2 K)y_0 + \tau \dot{y}_0$$
$$\dot{y}_1 = \dot{y}_0 - \tfrac{1}{2}\tau K(y_0 + y_1) \qquad (9.92)$$

and the matrix $I + \tfrac{1}{4}\tau^2 K$ has to be inverted, or factored, in order to pass from time level 0 at t to time level 1 at $t + \tau$.

No higher-order single step method is known at the present time which is at once unconditionally stable and conserving, and hence the practical importance of the scheme (9.89).

If the vibrating system itself is not conserving, the stable numerical integration scheme will generate or dissipate energy accordingly. Whether energy is supplied or subtracted from the system might very well not be obvious from the differential equation. Consider for example the pendulum in Fig. 9.9, driven by a mechanism that periodically changes its length l. This is a well known case of *parametric excitation* and the pertinent equation of motion is the *Mathieu* equation

$$\ddot{y} + (a - 2q \cos 2t)y = 0 \qquad (9.93)$$

where a and q are related to the parameters of the pendulum and the driving mechanism. This equation is linear but with time dependent coefficients. Newmark's method is used to solve this equation with $\tau = \pi/40$, and for the

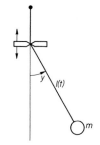

Fig. 9.9   Parametrically excited simple pendulum.

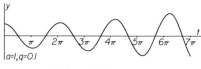

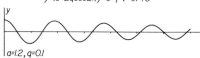

Fig. 9.10   Solution of Mathieu's equation $\ddot{y} + (a - 2q \cos 2t)y = 0$ for different values of the parameters a and q.

two sets of parameters, $a = 1$, $q = 0.1$ and $a = 1.2$, $q = 0.1$. In the first case, shown in Fig. 9.10, energy is pumped into the pendulum by the driving device, while in the second case, the motion is decaying because energy is extracted from the pendulum by the driving device.

9. Unconditionally Stable Semiexplicit Schemes

For the linear single degree of freedom dynamic problem, there exists no fundamental *computational* distinction between explicit and implicit schemes applied to its solution, and the step size limitation of stability is beyond that dictated by the desire for reasonable accuracy. Serious computational differences between explicit conditionally stable and implicit unconditionally stable schemes arise only in nonlinear, say single degree of freedom, problems and those leading to a stiff *system* of linear coupled equations of motion such as typically occur in the elastodynamic problem discretized by finite differences. In nonlinear problems, implicit schemes require the solution of a nonlinear equation at each step while the explicit scheme does not.

Solution of $Ky + \ddot{y} = 0$ by an explicit scheme imposes the stability condition that the step size τ be a fraction of the *shortest* natural period in the system, that of the highest mode, even if the actual motion of the elastic system is composed of only the few lowest modes. Implicit schemes are free of this restriction but require the solution of each step of a system of equations with a matrix such as $I + \frac{1}{4}\tau^2 K$ of Eq. (9.92). If this matrix fits well into the computer core, it can be inverted once and used as long as τ does not change. Otherwise, if inversion of this matrix is prohibitive, it may be triangularly factored, with only the envelope stored, and two back substitutions performed at each step.

In this section, we present an intermediate procedure to solve $Ky + \ddot{y} = 0$ by a method that is unconditionally stable but which does not require the factoring of a τ-dependent matrix the size of K, but only back substitutions with triangular matrices that are derived from K simply by dividing it into sums. This is why we prefer to call this method semiexplicit. In the case of large sparse matrices with a complex pattern of nonzero entries, the computational difference between dividing the large matrix into a sum of triangular matrices and triangularly factoring it is, apart from the obvious additional labor needed for the factoring, that if K is divided into the *sum* of upper and lower triangular matrices, then their entries pattern is exactly the same as that of K, while factoring can bring about considerable *fill-ins* inside the envelope of the original matrix, requiring a reorganization of the storage arrangements for the big matrices.

But there is a price to pay for avoiding triangular factorization—the appearance of $(\tau/h^m)^2$, where $m = 1$ for the string and $m = 2$ for the beam, in the time integration error expression—when we assume the motion to consist of only the lowest few modes, and the spatial mesh size enters again into the time integration procedure, limiting τ by *accuracy* considerations to the size of h^m.

So we propose to solve the system $Ky + \ddot{y} = 0$, $y(0) = y_0$, $\dot{y}(0) = \dot{y}_0$ in which K is symmetric and positive definite with the scheme

$$\left(I + \frac{\tau^2}{8} L\right)\dot{y}_{1/2} = \left(I - \frac{\tau^2}{8} L\right)\dot{y}_0 - \frac{\tau}{2} K y_0$$

$$y_{1/2} = y_0 + \frac{\tau}{4}(\dot{y}_0 + \dot{y}_{1/2})$$

$$\left(I + \frac{\tau^2}{8} U\right)\dot{y}_1 = \left(I - \frac{\tau^2}{8} U\right)\dot{y}_{1/2} - \frac{\tau}{2} K y_{1/2}$$

(9.94)

$$y_1 = y_{1/2} + \frac{\tau}{4}(\dot{y}_{1/2} + \dot{y}_1)$$

where L and U are lower and upper triangular matrices such that

$$L + U = K \qquad\qquad (9.95)$$

and where the stations 0, $\frac{1}{2}$, and 1 are at $t = t_0$, $t = t_0 + \frac{1}{2}\tau$, and $t = t_0 + \tau$. Execution of the scheme (9.94) requires the solution of only triangular systems with the matrices $I + \frac{1}{8}\tau^2 L$ and $I + \frac{1}{8}\tau^2 U$. Elimination of $y_{1/2}$ and $\dot{y}_{1/2}$ from the system of Eqs. (9.94) results in the direct passage from y_0 and $\dot{y}_0$ to y_1 and $\dot{y}_1$ through

$$\begin{bmatrix} I - \dfrac{\tau^2}{8} L & \dfrac{\tau}{2} K \\[2ex] -\dfrac{\tau}{2} I & I - \dfrac{\tau^2}{8} U \end{bmatrix} \begin{bmatrix} \dot{y}_1 \\[2ex] y_1 \end{bmatrix} = \begin{bmatrix} I - \dfrac{\tau^2}{8} L & -\dfrac{\tau}{2} K \\[2ex] \dfrac{\tau}{2} I & I - \dfrac{\tau^2}{8} U \end{bmatrix} \begin{bmatrix} \dot{y}_0 \\[2ex] y_0 \end{bmatrix} \qquad (9.96)$$

To prove the unconditional stability of the scheme (9.96), we shall show that if $U = L^T$, then the eigenvalues λ of

$$\begin{bmatrix} A & \dfrac{\tau}{2} K \\[2ex] -\dfrac{\tau}{2} I & B \end{bmatrix} \begin{bmatrix} x \\[2ex] y \end{bmatrix} = \lambda \begin{bmatrix} A & -\dfrac{\tau}{2} K \\[2ex] \dfrac{\tau}{2} I & B \end{bmatrix} \begin{bmatrix} x \\[2ex] y \end{bmatrix} \qquad (9.97)$$

where $A = I - \frac{1}{8}\tau^2 L$ and $B = I - \frac{1}{8}\tau^2 U$ are all of modulus 1 independently of τ. In fact, elimination of x in Eq. (9.97) leaves us with

$$ABy = -[\tfrac{1}{2}\tau(1 - \lambda)/(1 + \lambda)]^2 Ky \tag{9.98}$$

and if $U = L^T$, then $B = A^T$, AB becomes symmetric and at least positive semidefinite, and consequently

$$[(\tfrac{1}{2}\tau(1 - \lambda)/(1 + \lambda)]^2 = -\alpha^2(\tau) \tag{9.99}$$

where $ABy = \alpha^2(\tau)Ky$. It readily results from Eq. (9.99) that $|\lambda| = 1$.

To study the accuracy of the scheme (9.96), we proceed to find the residuals r_1 and r_2 of the two equations by expanding y_1 and $\dot{y}_1$ around point 0, and have that

$$\begin{aligned} r_1 &= -\tfrac{1}{12}\tau^3 \dddot{y}_0 - \tfrac{1}{8}\tau^3 L \ddot{y}_0 \\ r_2 &= -\tfrac{1}{12}\tau^3 \dddot{y}_0 - \tfrac{1}{8}\tau^3 U \dot{y}_0 \end{aligned} \tag{9.100}$$

We are interested in seeing what r_1 and r_2 become when the motion is spanned by only the first few modes and assume therefore that $y = x_1 e^{i\omega t}$, where ω is the lowest natural frequency of the system and x_1 is such that $Kx_1 = \omega^2 x_1$, which solves $\ddot{y} + Ky = 0$. Now $\|y\| = \|x_1\| = 1$, $\|\dot{y}\| = \omega$, $\|\ddot{y}\| = \omega^2$, $\|\dddot{y}\| = \omega^3$, and $\|\ddddot{y}\| = \omega^4$. But because x_1 is not necessarily an eigenvector of L or U, then $Lx_1 = O(h^{-2m})$ because in problems of order $2m$, $\|K\| = O(h^{-2m})$. Hence,

$$\begin{aligned} r_1 &= O(\tau^3) + O(\tau^3/h^{2m}) \\ r_2 &= O(\tau^3) + O(\tau^3/h^{2m}) \end{aligned} \tag{9.101}$$

and after some time we expect the error in the solution to be $O(\tau^2/h^{2m})$.

To demonstrate numerically the peculiarities of the scheme (9.94), we choose the simple problem of a fixed unit string discretized with linear finite elements and a lumped mass matrix. The string is released from rest in the shape of its first natural mode $y(x, 0) = \sin(\pi x)$ and is left to freely move according to $y(x, t) = \sin(\pi x) \cos(\pi t)$. First we discretize the string with 20 elements, i.e., $h = 0.05$. Had we chosen to solve this problem with the explicit scheme (9.33), we would have been restricted by stability to $\tau < 0.05$, and this is why we choose $\tau = 0.05$ in Eq. (9.94). The result of this computation is shown in Fig. 9.11 and seems adequately correct. Next we divide the string into a finer mesh of 200 elements, i.e., $h = 0.005$, and reapply the scheme (9.94) with the same time step $\tau = 0.05$. Figure 9.11 shows that the computed results are now in great error. Granted that if the numerical results with $h = 0.05$ are good enough, then $h = 0.005$ is an overdiscretization, but the point is that the scheme (9.94) forces us to overdiscretize now also in time to maintain the previous accuracy.

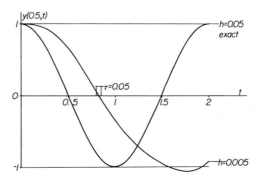

Fig. 9.11 Computed motion of the center point of a unit fixed string, started from rest, in the shape of the first eigenvector of the global stiffness matrix. The *increased* error introduced by scheme (9.94) with *reduced* mesh size is evident.

Does the error in y increase proportionately to $1/h^2$? With $\tau = 0.05$, at $t = 1.5$, the error in y is 1.17% when $h = \frac{1}{20}$, but is 94% when $h = \frac{1}{200}$, increasing by about a factor of 100 when h is decreased by a factor of 10.

10. Multistep Methods

A general *two step* method, correct for $x = 1$, $x = t$, and $x = t^2$, can be written as

$$y_2 = -y_0 + 2y_1 + \tau^2(\beta_0\ddot{y}_0 + \beta_1\ddot{y}_1 + \beta_2\ddot{y}_2), \qquad \beta_0 + \beta_1 + \beta_2 = 1 \qquad (9.102)$$

in which τ is the time step between levels 0 and 1, as well as between time levels 1 and 2. For the linear one variable equation $\ddot{y} + \omega^2 y = 0$, the scheme (9.102) becomes

$$y_2 = -y_0 + 2y_1 - \psi(\beta_0 y_0 + \beta_1 y_1 + \beta_2 y_2), \qquad \beta_0 + \beta_1 + \beta_2 = 1 \qquad (9.103)$$

and if we choose $\beta_0 = 0$, $\beta_2 = 0$, $\beta_1 = 1$, it becomes

$$y_0 - 2y_1 + y_2 + \psi y_1 = 0 \qquad (9.104)$$

which is a *central* scheme for y at point 1. The characteristic equation of this scheme is

$$z^2 - 2z(1 - \tfrac{1}{2}\psi) + 1 = 0, \qquad \psi = \omega^2\tau^2 \qquad (9.105)$$

and for z to be complex, we need to have

$$|1 - \tfrac{1}{2}\psi| < 1, \qquad 0 < \tau < 2/\omega \qquad (9.106)$$

Since also $|z| = 1$, the scheme

$$y_2 = -y_0 + y_1(2 - \psi), \qquad \psi = \omega^2\tau^2 \qquad (9.107)$$

to predict y_2 from the *two* previous values y_0 and y_1 is explicit, conserving, and stable when $\tau < 2/\omega$. The characteristic equation, which actually characterizes the scheme, of the central scheme (9.107) is the same as that of the conditionally stable single step method (9.33), and hence the two methods are equivalent. But to start the two step scheme (9.107), we need two *displacement* values with one computed from the initial velocity; the two step scheme is not self-starting.

Leaving β_0, β_1, and β_2 in Eq. (9.102), its characteristic equation becomes

$$z^2 - 2z\frac{1 - \frac{1}{2}\psi\beta_1}{1 + \psi\beta_2} + \frac{1 + \psi\beta_0}{1 + \psi\beta_2} = 0 \qquad (9.108)$$

To make z complex and of modulus 1, we choose $\beta_0 = \frac{1}{4}$, $\beta_1 = \frac{1}{2}$, and $\beta_2 = \frac{1}{4}$, with which the scheme (9.103) becomes implicit, with the characteristic equation

$$z^2 - 2z\frac{1 - \frac{1}{4}\psi}{1 + \frac{1}{4}\psi} + 1 = 0 \qquad (9.109)$$

which is the same as that of Newmark's method. The implicit two step method

$$y_2 = -y_0 + 2y_1 + \tfrac{1}{4}\tau^2(\ddot{y}_0 + 2\ddot{y}_1 + \ddot{y}_2) \qquad (9.110)$$

is thus equivalent to the implicit one step method (9.89).

One of the earliest *three* step methods is due to *Houbolt*, (and we mention this scheme for historical reasons—it being of no particular interest otherwise), who suggested the difference scheme

$$\ddot{y}_3 = (1/\tau^2)(2y_3 - 5y_2 + 4y_1 - y_0) \qquad (9.111)$$

which for $\ddot{y} + \omega^2 y = 0$ becomes

$$y_3(2 + \psi) - 5y_2 + 4y_1 - y_0 = 0 \qquad (9.112)$$

associated with the cubic characteristic equation

$$z^3(2 + \psi) - 5z^2 + 4z - 1 = 0, \qquad \psi = \omega^2\tau^2 \qquad (9.113)$$

It posesses one real root r, shown in Fig. 9.12, and two complex conjugate roots. The real root r is seen in Fig. 9.12 to monotonically decrease from $r = \frac{1}{2}$ at $\psi = 0$ to $r = 0$ at $\psi = \infty$. Once this real root is available, the cubic equation (9.105) can be reduced to the quadratic

$$z^2 + z\frac{1 - 4r}{r^2(2 + \psi)} + \frac{1}{r(\psi + 2)} = 0 \qquad (9.114)$$

The discriminant Δ of this equation is

$$\Delta = (-4r^2 + 8r - 3)/r^4(2 + \psi)^2 \qquad (9.115)$$

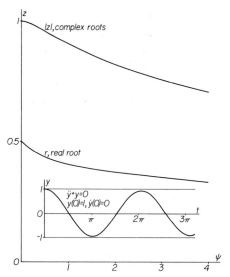

Fig. 9.12   Real and complex roots of scheme (9.112) and its application to $\ddot{y} + y = 0$, $y(0) = 1$, $\dot{y}(0) = 0$.

and is negative when $r < \frac{1}{2}$. Hence, the two other roots are complex for all $\psi > 0$, and $|z|$ is shown in Fig. 9.12. Houbolt's scheme is unconditionally stable in the sense that no energy is generated by it, but it is *dissipating* since $|z| < 1$ when $\psi > 0$. The numerical results for $y + \ddot{y} = 0$, $y(0) = 1$, $\dot{y}(0) = 0$, undertaken with Houbolt's scheme for $\tau = \pi/10$, are shown in Fig. 9.11 and the loss of amplitude is noticeable.

The solution to Houbolt's scheme may be written as

$$y_j = c_1 r^j + |z|^j (c_2 \cos \theta + c_3 \sin \theta) \tag{9.116}$$

Assume $\psi = 0.1$, and suppose that y_0, y_1, and y_2 are exactly available from the solution $y = \cos \omega t$. Then

$$y_j = c_1 e^{-2.32\omega t} + e^{-0.0095\omega t}(c_2 \sin 0.961\omega t + c_3 \cos 0.961\omega t) \tag{9.117}$$

and if $\omega^2 = 1$

$$c_1 = 0.025, \qquad c_2 = -0.0426, \qquad c_3 = 1.0168 \tag{9.118}$$

In order to get complex roots of modulus 1 from a cubic characteristic equation associated with a three step method, the equation ought to be of the form

$$z^3 + az^2 + az + 1 = 0 \tag{9.119}$$

Then, its real root is -1, while the complex ones are obtained from the reduced quadratic

$$z^2 + z(a - 1) + 1 = 0 \qquad (9.120)$$

in which necessarily $|a - 1| < 2$.

A characteristic equation of the form (9.119) is associated with the *symmetric* or central three step scheme:

$$0 = -y_0 + \alpha_1 y_1 + \alpha_1 y_2 - y_3 + \tau^2(\beta_0 \ddot{y}_0 + \beta_1 \ddot{y}_1 + \beta_1 \ddot{y}_2 + \beta_0 \ddot{y}_3) \qquad (9.121)$$

which, when made exact for $y = 1$, $y = t$, $y = t^2$, and $y = t^3$, becomes

$$0 = -y_0 + y_1 + y_2 - y_3 + \tau^2[\beta_0 \ddot{y}_0 + (1 - \beta_0)\ddot{y}_1 + (1 - \beta_0)\ddot{y}_2 + \beta_0 \ddot{y}_3] \qquad (9.122)$$

having the characteristic equation

$$z^3 - z^2 \frac{1 - \psi(1 - \beta_0)}{1 + \beta_0 \psi} - z \frac{1 - \psi(1 - \beta_0)}{1 + \beta_0 \psi} + 1 = 0 \qquad (9.123)$$

which is indeed of the form (9.117). It is solved by $z = -1$ and then reduces to the quadratic

$$z^2 - 2z \frac{1 - \psi(\frac{1}{2} - \beta_0)}{1 + \beta_0 \psi} + 1 = 0 \qquad (9.124)$$

from which the complex roots are obtained. They are complex conjugate when

$$\psi(1 - 4\beta_0) < 4 \qquad (9.125)$$

which holds for $\psi > 0$ when $\beta_0 \geq \frac{1}{4}$. For $\beta_0 = \frac{1}{4}$, the characteristic equation (9.124) becomes the same as (9.109).

Figure 9.13 shows the solution of $\ddot{y} + y = 0$, $y(0) = 1$, $\dot{y}(0) = 0$, undertaken using the symmetric scheme (9.122) with $\beta_0 = \frac{1}{4}$ and $\beta_0 = 1$.

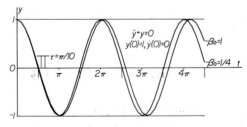

Fig. 9.13   Application of scheme (9.122) with different values of β_0 to $\ddot{y} + y = 0$, $y(0) = 1$, $\dot{y}(0) = 0$.

11. Runge–Kutta–Nyström Method

Corresponding to the Runge–Kutta method of Chapter 8, Section 17, for the first-order initial value problem is the Nyström *explicit* scheme

$$y_1 = y_0 + \tau \dot{y}_0 + \tau^2 \sum_{j=1}^{n} v_j k_j$$

$$\dot{y}_1 = \dot{y}_0 + \tau \sum_{j=1}^{n} w_j k_j \qquad (9.126)$$

$$k_i = f\left(y_0 + \alpha_i \tau \dot{y}_0 + \tau^2 \sum_{j=1}^{i-1} \beta_j k_j\right), \qquad i = 1, 2, \ldots, n$$

for the second-order equation $\ddot{y} = f(y)$, which has the advantage of circumventing the repeated differentiation of $f(y)$ in higher-order schemes, replacing it by the repeated evaluation of $f(y)$.

As an example of how the coefficients of Nyström's scheme are evaluated, assume $n = 1$. Then Eq. (9.126) becomes

$$y_1 = y_0 + \tau \dot{y}_0 + \tau^2 v_1 k_1$$

$$\dot{y}_1 = \dot{y}_0 + \tau w_1 k_1 \qquad (9.127)$$

$$k_1 = f(y_0 + \alpha_1 \tau \dot{y}_0)$$

Assume further that $f(y) = -y$, as in the linear case, and you get that

$$y_1 = (1 - \tau^2 v_1) y_0 + \tau (1 - \tau^2 \alpha_1 v_1) \dot{y}_0$$

$$\dot{y}_1 = -\tau w_1 y_0 + (1 - \tau^2 \alpha_1 w_1) \dot{y}_0 \qquad (9.128)$$

By choosing $v_1 = \frac{1}{2}$, $\alpha_1 = \frac{1}{2}$, and $w_1 = 1$ this becomes the same as the scheme (9.45), and hence the time integration method

$$y_1 = y_0 + \tau \dot{y}_0 + \tfrac{1}{2} \tau^2 k_1$$

$$\dot{y}_1 = \dot{y}_0 + \tau k_1 \qquad (9.129)$$

$$k_1 = f(y_0 + \tfrac{1}{2} \tau \dot{y}_0)$$

is conditionally stable, conserving, and with accuracy $O(\tau^2)$ in the frequency.

12. Shooting in Boundary Value Problems

Shooting is the name given in the literature to *initial* value methods for the solution of *boundary* value problems as a metaphor to the gunner's practice of correcting his aim according to the previous hit. In essence, it is a conceptually rather simple method that is readily explained with reference

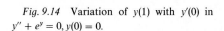

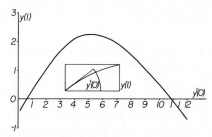

Fig. 9.14 Variation of $y(1)$ with $y'(0)$ in $y'' + e^y = 0, y(0) = 0.$

to Fig. 9.14. Suppose that we are interested in solving the initial value problem $-y'' = f(y)$, $0 < x < 1$; $y(0) = y_0$, $y(1) = y_1$. Instead, we replace it by the initial value problem $-y'' = f(y)$, $y(0) = y_0$, and $y'(0) = \alpha_1$, where α_1 is the first guess for $y'(0)$, and proceed to compute $y(1)$ by one of the techniques described in this chapter. Assume that this trial shot hits the wall $x = 1$ at $y(1) = \beta_1$. Normally $\beta_1 \neq y_1$, and we shoot again with a different initial angle, say α_2, this time hitting the wall $x = 1$ at $y(1) = \beta_2$. Assuming a linear relationship between α and β, we have that

$$\beta = \frac{\alpha_2 \beta_1 - \alpha_1 \beta_2}{\alpha_2 - \alpha_1} + \frac{\beta_2 - \beta_1}{\alpha_2 - \alpha_1} \alpha \tag{9.130}$$

and the next shot is fired at an angle α corresponding to $\beta = y_1$ in Eq. (9.130). Usually this will still be off the mark and the process is repeated, with various degrees of program sophistication, to a hopeful convergence.

Shooting works best when the dependence of β upon α is continuous and monotonic. Otherwise, the shooter stands the risk of wasting his ammunition.

Shooting might help us discover solutions to the nonlinear boundary value problem, to which successive substitutions do not converge. Take, for example, the boundary value problem

$$y'' + e^y = 0, \qquad 0 < x < 1$$
$$y(0) = y(1) = 0 \tag{9.131}$$

Shooting in different initial angles $y'(1)$, we discover in Fig. 9.14 that this nonlinear problem has *two* solutions, as shown in Fig. 9.15.

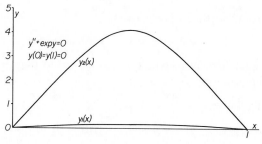

Fig. 9.15 Two Solutions of $y'' + e^y = 0$, $y(0) = y(1) = 0$.

EXERCISES

1. In the presence of damping, the equation of motion of a linear elastic
    system becomes

    $$Ky + C\dot{y} + M\ddot{y} = 0$$

    where K and M are the stiffness and mass matrices and C is the *damping
    matrix*. Suppose that a linear transformation reduces this equation to

    $$Ky + C\dot{y} + \ddot{y} = 0$$

    Let X be the matrix of eigenvector columns of $Kx = \lambda x$ and Λ the
    diagonal matrix $X^T K X$. Show that $y = X_z$ transforms the last equation
    into

    $$\Lambda z + X^T C X z + \ddot{z} = 0$$

    when $X^T X = I$.
        Decoupling of the last equation occurs with $y = Xz$ if C is diagonal-
    ized by $X^T C X$, or when C and K have the same eigenvector system. A
    necessary and sufficient condition for this is that K and C commute.
    Show that the following choices for C, $C = \alpha K + \beta K^{-1}$ or $C = \alpha K +
    \beta K^2$, for instance, result in decoupling.

2. Show that the characteristic equation of the *four* step scheme

    $$12\tau^2 \ddot{y}_2 = -y_0 + 16y_1 - 30y_2 + 16y_3 - y_4$$

    is

    $$z^4 - 16z^3 + 6z^2(5 - 2\psi) - 16z + 1 = 0$$

    Explore the stability properties of this scheme.

3. Solve the nonlinear equation of motion

    $$\ddot{y} + y^3 = 0, \qquad y(0) = 1, \qquad \dot{y}(0) = 0$$

    with an appropriate scheme from this chapter. Pay attention to the
    round-off and discretization errors involved.

4. Numerically solve *Ricatti's* equation

    $$\dot{y} + y^2 = 1, \qquad y(0) = 0$$

    This nonlinear first-order equation is reduced to the linear $\ddot{u} - u = 0$
    by the transformation $y = \dot{u}/u$.

5. Numerically solve

    $$\ddot{y} = 6y^2, \qquad -1 < t < 1, \qquad y(0) = 1, \qquad \dot{y}(0) = 0$$

and

$$\ddot{y} = 2y^3 + ty, \qquad y(0) = 1, \qquad \dot{y}(0) = 0$$

6. Solve the equation of motion of the pendulum

$$\ddot{y} + \sin y = 0, \qquad y(0) = \pi/2, \qquad \dot{y}(0) = 0$$

7. If the axis of the pendulum rotates with angular velocity ω, its equation of motion becomes

$$\ddot{y} + \sin y - \tfrac{1}{2}\omega^2 \sin 2y = 0$$

Solve it with the initial conditions $y(0) = \pi/4$, $\dot{y}(0) = 0$ for values of ω ranging from 0 to 10 and describe the motion.

8. Solve the equation of motion $\ddot{y} + f(y) = 0$, $y(0) = 2$, $\dot{y}(0) = 0$, where the restoring force is as in Fig. 9.16.

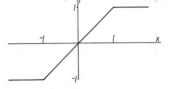

Fig. 9.16   Force–displacement relationship.

9. Solve *Duffing's* equation of motion

$$\ddot{y} + 0.7\dot{y} + y^3 = 0.75 \cos t, \qquad y(0) = \dot{y}(0) = 0$$

10. Solve Duffing's equation

$$\ddot{y} + 0.2\dot{y} + y|y| = 1.5 \cos 2t + 0.5, \qquad y(0) = \dot{y}(0) = 0$$

and compare your results with Fig. 9.17.

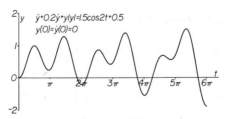

Fig. 9.17   Duffing's equation, $\ddot{y} + 0.2\dot{y} + y|y| = 1.5 \cos 2t + 0.5$, $y(0) = \dot{y}(0) = 0$.

11. Solve Duffing's equation

$$\ddot{y} + y - \tfrac{1}{6}y^3 = 0.8 \sin 0.27\omega t, \qquad \omega = 0.92845, \qquad y(0) = \dot{y}(0) = 0$$

and compare your results with Fig. 9.18. Observe the sudden jump phenomenon.

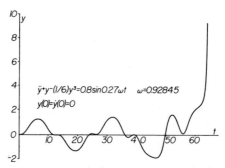

Fig.9.18   Duffing's equation, $\ddot{y} + y - \frac{1}{6}y^3 = 0.8 \sin 0.27\omega t$, $\omega = 0.92845$, $y(0) = \dot{y}(0) = 0$.

12.   Solve *Van der Pol*'s equation

$$\ddot{y} - (1 - y^2)\dot{y} + y = 0, \qquad y(0) = 0, \qquad \dot{y}(0) = -0.05$$

and compare your result with Fig. 9.19.

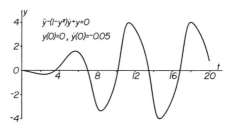

Fig. 9.19   Van der Pol's equation, $\ddot{y} - (1 - y^2)\dot{y} + y = 0$, $y(0) = 0$, $\dot{y}(0) = -0.05$.

13.   Solve Van der Pol's equation

$$\ddot{y} - 5(1 - y^2)\dot{y} + y = 0, \qquad y(0) = 2, \qquad \dot{y}(0) = 0$$

and compare your result with Fig. 9.20.

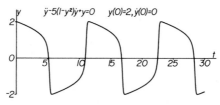

Fig. 9.20   Van der Pol's equation, $\ddot{y} - 5(1 - y^2)\dot{y} + y = 0$, $y(0) = 2$, $y(0) = 0$.

14.   Solve the equation of motion with quadratic damping

$$\ddot{y} + \frac{1}{2}\dot{y}^2 + y = 0, \qquad y(0) = 1, \qquad \dot{y}(0) = 0$$

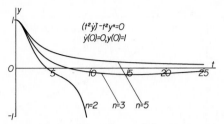

Fig. 9.21 Emden's equation, $(t^2\dot{y})' - t^2 y^2 = 0$, $\dot{y}(0) = 0$, $y(0) = 0$.

15. Solve *Emden's* equation

$$(t^2\dot{y})' + t^2 y^n = 0, \qquad n = 2, 3, 5, \qquad y(0) = 1, \qquad \dot{y}(0) = 0$$

and compare your results with Fig. 9.21.

16. Solve Emden's equations

$$(t^2\dot{y})'/t^2 + f(y) = 0, \qquad y(0) = 1, \qquad \dot{y}(0) = 0$$

where

      (i) $f(y) = \sin y$

      (ii) $f(y) = \cos y$

and compare your results with Figs. 9.22 and 9.23.

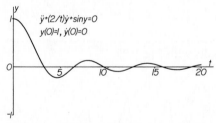

Fig. 9.22 Emden's equation, $\ddot{y} + (2/t)\dot{y} + \sin y = 0$, $y(0) = 1$, $\dot{y}(0) = 0$.

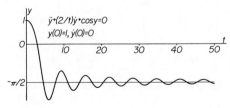

Fig. 9.23 Emden's equation, $\ddot{y} + (2/t)\dot{y} + \cos y = 0$, $y(0) = 1$, $\dot{y}(0) = 0$.

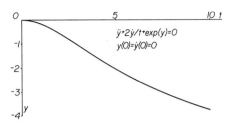

Fig. 9.24   Liouville's equation, $y + 2\dot{y}/t + e^y = 0$, $y(0) = \dot{y}(0) = 0$.

17.   Solve *Liouville's* equation

$$(t^2 \dot{y})^{\cdot} + t^2 e^y = 0, \qquad y(0) = \dot{y}(0) = 0$$

and compare your result with Fig. 9.24.

18.   Let x and y be the members of two competing species. Their ecological equilibrium is described by *Volterra's* set of equations

$$\dot{x} = ax(1 - y), \qquad \dot{y} = -cy(1 - x)$$

or

$$x\ddot{x} = \dot{x}^2 + acx^2 - cx\dot{x} + cx^2\dot{x} - acx^3$$

$$y\ddot{y} = \dot{y}^2 + acy^2 + ay\dot{y} - ay^2\dot{y} - acy^3$$

Solve these equations for $a = 2$, $c = 1$, and with the initial conditions $x(0) = 1$, $y(0) = 3$. Make use of the first integral E:

$$E = cx + ay - c\log x - a\log y$$

to check the conservation of the numerical scheme you use and compare your results with Fig. 9.25.

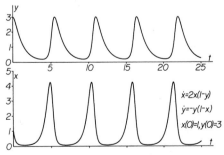

Fig. 9.25   Volterra's equation of competing species, $\dot{x} = 2x(1 - y)$, $\dot{y} = -y(1 - x)$, $x(0) = 1$, $y(0) = 3$.

19. The equations of motion of a point mass m, centrally attracted by a force $f(r)$, are given in polar coordinates by

$$m(\ddot{r} - r\dot{\phi}^2) = f(r)$$

$$r\ddot{\phi} + 2\dot{r}\dot{\phi} = \frac{1}{r}(r^2\dot{\phi})^{\cdot} = 0$$

Associated with this motion, when $f(r) = -m\gamma r^{-n}$, are the two invariants or first integrals of motion

$$E = \frac{1}{2}(\dot{r}^2 + r^2\dot{\phi}^2) - \frac{m\gamma}{r^{n-1}}\frac{1}{n-1}, \qquad n > 1$$

and

$$H = mr^2\dot{\phi}$$

expressing conservation of energy and *angular momentum*, respectively. Solve this pair of equations for $\gamma = m = 1$, $n = 2$, $r(0) = 1$ $\dot{r}(0) = 0$, $\phi(0) = 0$, $\dot{\phi}(0) = 1.01$. Had we taken as initial condition $\dot{\phi}(0) = 1.0$ instead of $\dot{\phi}(0) = 1.01$, the orbit described by the mass m would have been circular. Show that here it is nearly circular.

20. Repeat Exercise (19) but with $n = 4$ and explain the resulting motion in terms of the *orbital instability* that occurs when $n > 3$.

21. Consider the pendulum suspended on an elastic string, shown in Fig. 9.26, that may swing both in the ϕ direction and in the y direction along the string. Assuming that this motion is such that $\cos\phi = 1 - \frac{1}{2}\phi^2$, the equations describing it become

$$\ddot{y} + \omega_2^2 y + \frac{1}{2}\omega_1^2\phi^2 - \dot{\phi}^2 = 0$$

$$\ddot{\phi} + \omega_1^2\phi + \omega_1^2 y\phi + 2\dot{y}\dot{\phi} + 2y\ddot{\phi} = 0$$

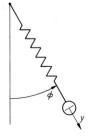

Fig. 9.26 Pendulum suspended on a spring.

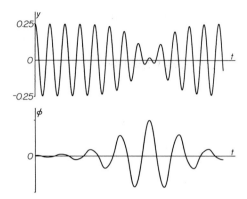

Fig. 9.27   Motion of the pendulum in Fig. 9.26.

where $\omega_2{}^2 = k/m$ and $\omega_1{}^2 = g/l$, g being the gravitational acceleration, l the length of the string when $y = 0$, and y the non dimensional displacement relative to l. This pair of equations is characterized by a *nonlinear coupling*, since their linearization would have decoupled them. The potential plus kinetic energy of the pendulum is proportional to E:

$$E = \dot{y}^2 + \dot{\phi}^2 + 2y\dot{\phi}^2 + \omega_2{}^2 y^2 + \omega_1{}^2 \phi^2 + \omega_1{}^2 y\phi^2$$

Solve the equations of motion for $\omega_1 = 1$, $\omega_2 = 2$, with the initial conditions $y(0) = 0.25$, $\dot{y}(0) = 0$, $\phi(0) = 0.001$, $\dot{\phi}(0) = 0$, and compare your results with Fig. 9.27. Use the first integral of the motion E to check the conservation of scheme you use. The instability of the motion in the ϕ direction—the transfer of energy from the y to the ϕ mode of vibration is termed *autoparametric excitation*.

22.   Let a point P move with a constant velocity v in a circle of radius r. Another point Q, at a distance y from P, pursues P, chasing it along the line PQ, with a velocity kv, as shown in Fig. 9.28.

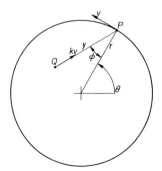

Fig. 9.28   Pursuit in a circle.

The differential equations describing this problem of pursuit in a circle are

$$y\frac{d\phi}{d\theta} = r\cos\phi - y$$

$$\frac{dy}{d\theta} = r(\sin\phi - k)$$

or when y is eliminated between them

$$\ddot{\phi}\cos\phi + (3\sin\phi - 2k)\dot{\phi} + (2\sin\phi - k)\dot{\phi}^2 + \sin\phi - k = 0$$

Solve this pair of equations for $r = 1$ and $k = \frac{2}{3}$, with the initial conditions $\phi(0) = 0$, $y(0) = 0.9999$; $\phi(0) = 0$, $y(0) = 2$; $\phi(0) = 0$, $y(0) = 3$, and compare your results with Figs. 9.29–9.31. Observe that whatever the initial conditions the pursuer finishes by moving on the *limit cycle* which is in this case an interior concentric circle of a radius equal to k.

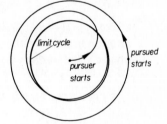

Fig. 9.29 Pursuit in a circle. Initial conditions: $\phi(0) = 0$, $y(0) = 0.9999$.

Fig. 9.30 Pursuit in a circle. Initial conditions: $\phi(0) = 0$, $y(0) = 2$.

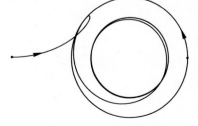

Fig. 9.31 Pursuit in a circle. Initial conditions: $\phi(0) = 0$, $y(0) = 3$.

23. When Newmark's (also *average acceleration*) method of Eq. (9.89) is applied to the nonlinear vibration problem $\ddot{y} + f(y) = 0$ it produces the implicit scheme:

$$y_1 + \tfrac{1}{4}\tau^2 f(y_1) = y_0 + \tau \dot{y}_0 - \tfrac{1}{4}\tau^2 f(y_0)$$

$$\dot{y}_1 = \dot{y}_0 - \tfrac{1}{2}\tau[f(y_0) + f(y_1)]$$

Use this scheme to solve $\ddot{y} + 100 \tanh(y) = 0$, $y(0) = 0$, $\dot{y}(0) = 25$ (first integral $E = \tfrac{1}{2}\dot{y}^2 + 100 \log \cosh(y)$, period $= 1.117$, amplitude $= 3.818$). You will need the Newton–Raphson method to solve a transcendental equation in order to find y_1 from y_0 and $\dot{y}_0$. Show that when $\tau = 0.2$ the scheme becomes strongly generating and the solution becomes unstable.

But the above scheme can be written differently for nonlinear problems, namely

$$y_1 = y_0 + \tau \dot{y}_0 - \tfrac{1}{2}\tau^2 f(y_a)$$

$$\dot{y}_1 = y_0 - \tau f(y_b)$$

where $y_a = (1 - \alpha)y_0 + \alpha y_1$ and $y_b = (1 - \beta)y_0 + \beta y_1$. Apply this scheme with $\alpha = \tfrac{1}{2}$, $\beta = \tfrac{1}{2}$ to the above initial value problem and show that now when $\tau = 0.2$ stability is restored to the solution for a long period of time.

SUGGESTED FURTHER READING

Minorsky, N., *Nonlinear Oscillations.* Van Nostrand Reinhold Princeton, New Jersey, 1962.

Davis, H. T., *Introduction to Nonlinear Differential and Integral Equations.* Dover, New York, 1962.

Bathe, K. J., and Wilson, E. L., *Numerical Methods in Finite Element Analysis.* Prentice-Hall, Englewood Cliffs, New Jersey, 1976.

10 *Wave Propagation*

1. Standing and Traveling Waves in a String

In Chapter 6, Section 2, we developed the equation of motion of the stretched string fixed at both its ends, and had there that for a string of uniform properties, the displacement $y(x, t)$ is given by

$$c^2 \frac{\partial^2 y}{\partial x^2} = \frac{\partial^2 y}{\partial t^2}, \qquad c = \text{const.,} \quad 0 < x < l \qquad (10.1)$$

$$y(0, t) = 0, \qquad y(l, t) = 0, \quad t > 0 \qquad (10.2)$$

$$y(x, 0) = f(x), \qquad \dot{y}(x, 0) = g(x) \qquad (10.3)$$

Equation (10.1) is the *linear wave equation*, Eq. (10.2) expresses the boundary conditions, here fixed end points, while Eq. (10.3) describes the initial conditions of configuration and velocity.

Separation of variables is successful in the wave equation and we write the displacement $y(x, t)$ as the product of a function of x only and a function of t only in the form

$$y(x, t) = y(x)z(t) \qquad (10.4)$$

and have upon its introduction into the wave equation (10.2) that

$$\frac{y''}{y} = \frac{1}{c^2} \frac{\ddot{z}}{z} \qquad (10.5)$$

Inasmuch as y is a function of x only, and z a function of t only, Eq. (10.5) necessarily equals a constant, say $-\lambda$, and the wave equation separates into

$$y'' + \lambda y = 0, \qquad 0 < x < l \qquad (10.6)$$

$$\ddot{z} + \lambda c^2 z = 0, \qquad t > 0 \qquad (10.7)$$

where the boundary conditions for Eq. (10.6) are gotten from Eq. (10.2) through

$$y(0)z(t) = 0, \qquad y(l)z(t) = 0 \tag{10.8}$$

leading to

$$y(0) = y(l) = 0 \tag{10.9}$$

We recognize Eqs. (10.6) and (10.9) to be the fixed string eigenproblem, which we know possesses the nontrivial solutions

$$y_j(x) = \sin(\pi j x/l), \qquad j = 1, 2, 3, \ldots, \infty \tag{10.10}$$

and the corresponding eigenvalues

$$\lambda_j = \omega_j^2 = (\pi j/l)^2, \qquad j = 1, 2, 3, \ldots, \infty \tag{10.11}$$

Now that we have λ_j, we introduce it into Eq. (10.7), whose solution becomes

$$z_j(t) = a_j \cos \omega_j c t + b_j \sin \omega_j c t \tag{10.12}$$

where a_j and b_j are still undetermined coefficients. Each of the functions $y_j(x)z_j(t)$ is a solution of the wave equation as is their sum and hence

$$y(x, t) = \sum_{j=1}^{\infty} \sin \omega_j x (a_j \cos \omega_j c t + b_j \sin \omega_j c t) \tag{10.13}$$

Setting $t = 0$ in Eq. (10.13), we have that

$$y(x, 0) = \sum_{j=1}^{\infty} a_j \sin(\pi j x/l) = f(x)$$

$$\dot{y}(x, 0) = \sum_{j=1}^{\infty} b_j \omega_j c \sin(\pi j x/l) = g(x) \tag{10.14}$$

indicating that a_j and b_j are the Fourier coefficients of $f(x)$ and $g(x)$, the initial configuration and speed of the string. In the event of zero initial velocity, $\dot{y}(x, 0) = 0$, $b_j = 0$, and

$$y(x, t) = \sum_{j=1}^{\infty} a_j \sin \omega_j x \cos \omega_j c t, \qquad \omega_j = \pi j/l \tag{10.15}$$

With the trigonometric identity

$$\sin \alpha \cos \beta = \tfrac{1}{2}[\sin(\alpha + \beta) + \sin(\alpha - \beta)] \tag{10.16}$$

Eq. (10.15) becomes

$$y(x, t) = \sum_{j=1}^{\infty} \tfrac{1}{2} a_j [\sin \omega_j (x - ct) + \sin \omega_j (x + ct)] \tag{10.17}$$

and, from Eq. (10.14), we see that

$$\sum_{j=1}^{\infty} a_j \sin \omega_j(x - ct) = f(x - ct)$$

(10.18)

$$\sum_{j=1}^{\infty} a_j \sin \omega_j(x + ct) = f(x + ct)$$

where $f(x)$ is the initial wave configuration. Hence, $y(x, t)$ in Eq. (10.17) can be written as

$$y(x, t) = \tfrac{1}{2}[f(x + ct) + f(x - ct)]$$

(10.19)

in which $f(x)$ is *periodically extended* outside the string limits.

Equation (10.19) is the general solution to the wave equation when the waves start their travel from rest. According to Eq. (10.15) *all the Fourier components of $f(x)$ travel with the same speed c*, and $y(x, t)$ is half the sum of $f(x)$ traveling at velocity c to the left and to the right. Such wave propagation, in which the wave travels without change of profile, is termed *nondispersive*.

Examples of the solution (10.19) of the wave equations (10.1) are graphically shown in Figs. 10.1–10.3. Figure 10.1 shows a fixed string (in bold lines) initially plucked into the form of a symmetric triangle and released. To picture the string's shape as it varies with time, the initial shape of the string is periodically extended outside its limits, moved to the left and to the right with speed c, and averaged to obtain the standing wave form at different times. Figure 10.2 shows a triangular bump on a very long string, separating upon its release into two other triangular waves traveling in opposite

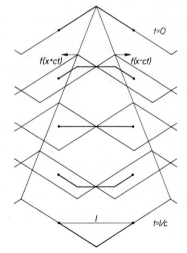

Fig. 10.1 Fixed string plucked at its center and released.

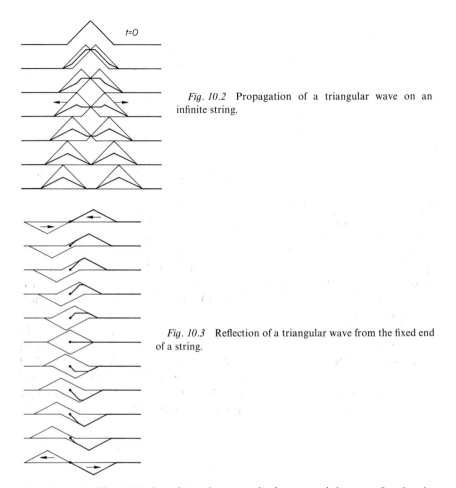

Fig. 10.2 Propagation of a triangular wave on an infinite string.

Fig. 10.3 Reflection of a triangular wave from the fixed end of a string.

directions. In Fig. 10.3, the triangular wave is shown arriving at a fixed point on the string and being reflected from it.

2.   Discretization in Space

The finite difference or finite element discretization in space of the wave equation is entirely analogous to that of the heat equation described in Chapter 8. In this, the acceleration $\ddot{y}$ is assumed to be a given function at any moment of time, and the wave equation

$$-y'' = -\ddot{y}/c^2 \tag{10.20}$$

is condidered in x only, with $-\ddot{y}/c^2$ looked upon as a load or source function, resulting in the coupled system of second-order ordinary differential equations

$$Ky + M\ddot{y} = 0 \tag{10.21}$$

in which K and M are the global stiffness and mass matrices, respectively, that we previously encountered in the finite element analysis of eigenproblems and heat transfer.

Equation (10.21) can be solved by modal decomposition, described in Chapter 9, Section 5, which requires the eigensystem of $Kx = \lambda Mx$, or it may be integrated in time stepwise with the aid of one of the algorithms presented in the previous chapter. In case a conditionally stable scheme is employed, say that in Eq. (9.33), the step size τ is restricted to

$$0 < \tau < 2/\max_x \{x^T Kx/x^T Mx\}^{1/2} \tag{10.22}$$

and the upper limit on τ can be estimated from the *element* eigenproblem $k_e y_e = \lambda^e m_e y_e$ as explained in Chapter 7, Section 1, and as was done in Chapter 8, Section 10, for the heat equation. For example, if the string is discretized with linear elements for which

$$k_e = \frac{1}{h}\begin{bmatrix} 1 & -1 \\ -1 & 1 \end{bmatrix}, \qquad m_e = \frac{h}{6}\begin{bmatrix} 2 & 1 \\ 1 & 2 \end{bmatrix} \tag{10.23}$$

then the largest element eigenvalue is $\lambda_2^{\,e} = 12h^{-2}$ and

$$0 < \tau < 0.58h \tag{10.24}$$

In the event that the string is discretized with a uniform mesh of cubic C^1 elements for which

$$k_e = \frac{1}{30h}\begin{bmatrix} 36 & 3 & -36 & 3 \\ 3 & 4 & -3 & -1 \\ -36 & -3 & 36 & -3 \\ 3 & -1 & -3 & 4 \end{bmatrix}, \qquad m_e = \frac{h}{420}\begin{bmatrix} 156 & 22 & 54 & -13 \\ 22 & 4 & 13 & -3 \\ 54 & 13 & 156 & -22 \\ -13 & -3 & -22 & 4 \end{bmatrix} \tag{10.25}$$

we obtain $\lambda_4^{\,e} = 170.12h^{-2}$ from the element eigenproblem as the largest eigenvalue and stability restricts τ in this case to

$$0 < \tau < 0.15h \tag{10.26}$$

In both cases the maximum allowed τ is $O(h)$, but for the same h the cubic element furnishes a better approximation.

3. Spurious Dispersion

A wave traveling on a string maintains its shape because the natural frequencies of the string are all integer multiples of π. The *computed* frequencies are not exact and as a result, the computed traveling wave deforms with time. To see more precisely how this happens, consider the string discretized with finite elements or finite differences that produce the system $Ky + M\ddot{y} = 0$, in which $y(t)$ is the vector containing the time dependent nodal values. Expansion of $y(t)$ in terms of the eigenvectors $x_1, x_2, \ldots, x_N$ of $Kx = \hat{\omega}^2 Mx$ in the form $y(t) = c_1(t)x_1 + \cdots + c_N(t)x_N$, decouples $Ky + M\ddot{y} = 0$, according to the analysis of Chapter 9, Section 5, into

$$\ddot{c}_j + \hat{\omega}_j{}^2 c_j = 0, \qquad j = 1, 2, \ldots, N \tag{10.27}$$

such that in the case of zero initial velocity

$$c_j(t) = c_j(0) \cos \hat{\omega}_j t \tag{10.28}$$

and

$$y(t) = \sum_{j=1}^{N} c_j(0) x_j \cos \hat{\omega}_j t \tag{10.29}$$

in which $c_1(0), c_2(0), \ldots, c_N(0)$ are obtained from the initial wave configuration. The computed frequencies $\hat{\omega}_j$ are all, to a different degree, disproportional to πj, and we write them as

$$\hat{\omega}_j = (\pi j c/l)(1 + \varepsilon_j) \tag{10.30}$$

Suppose now that we are engaged in numerically solving the string problem shown in Fig. 10.1. Because of the symmetry of the string and its initial triangular form, only the odd modes are included in the expression for $y(t)$. At $t = 0$, the initial triangular shape of the string is given by $y(0) = c_1(0)x_1 + c_3(0)x_3 + \cdots$. In order to get the inverted triangular shape at $t = 1/c$, $\cos \hat{\omega}_j t$ in Eq. (10.29) must be -1 for $j = 1, 3, 5, \ldots$, *all at the same time*. This happens when $\hat{\omega}_j = \omega_j = \pi j c/l$, but not for the discrete model, for which $\hat{\omega}_j$ is as given in Eq. (10.30). In the discrete case, $\cos \hat{\omega}_j t = -1$, $j = 1, 3, 5, \ldots$, occurs at

$$t = l/c(1 + \varepsilon_j) \tag{10.31}$$

which is different for different eigenvectors; the different eigenmodes in the eigenvector expansion for $y(t)$ travel at different speeds. Finite element discretization with a consistent mass matrix gives rise to $\varepsilon_j \geqslant 0$, and the higher modes travel faster than the lower ones. As a result of this difference in the speed of the different eigenmodes, the original shape of the wave deforms with time, with short wavelength modes, the ones that travel fastest, showing up. This artificial frequency dependent velocity of propagation of the different eigenmodes is termed *artificial*, *spurious*, or *numerical dispersion*.

Spurious dispersion does not originate in the space discretization alone. When $Ky + M\ddot{y} = 0$ is integrated stepwise, the time scheme introduces errors in the frequency too, as we have already observed in Chapter 9. If the explicit scheme

$$y_1 = y_0 + \tau \dot{y}_0 + \tfrac{1}{2}\tau^2 \ddot{y}_0$$
$$\dot{y}_1 = \dot{y}_0 + \tfrac{1}{2}\tau(\ddot{y}_0 + \ddot{y}_1)$$
(10.32)

is called to solve $\ddot{y} + \omega^2 y = 0$, $y(0) = y_0$, $\dot{y}(0) = 0$, then it follows from Eq. (9.36) that at the jth time step the computed value of y, y_j, is

$$y_j = y_0 \cos(\theta/\tau)t$$
(10.33)

in which θ/τ is the computed frequency given by

$$\sin^2 \theta = \psi(1 - \tfrac{1}{4}\psi), \qquad \psi = \omega^2 \tau^2$$
(10.34)

For the implicit scheme

$$y_1 = y_0 + \tau \dot{y}_0 + \tfrac{1}{4}\tau^2(\ddot{y}_0 + \ddot{y}_1)$$
$$\dot{y}_1 = \dot{y}_0 + \tfrac{1}{2}\tau(\ddot{y}_0 + \ddot{y}_1)$$
(10.35)

θ is given by

$$\sin^2 \theta = \psi/(1 + \tfrac{1}{4}\psi)^2$$
(10.36)

The relative error in the frequency $\theta/\omega\tau$, caused by the two schemes (10.32) and (10.35), is graphed in Fig. 10.4 where it is seen that for $\psi < 2.15$ the explicit scheme *increases* the frequency, while the implicit scheme *decreases* it. From Eq. (10.31), we conclude that the dispersion caused by the explicit time scheme is making the short wavelength modes advance faster since $\varepsilon_j \geq 0$ in this case, while the implicit scheme retards them.

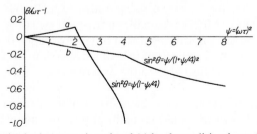

Fig. 10.4 Error in the frequency introduced (a) by the explicit scheme (10.32), (b) and the implicit scheme (10.35).

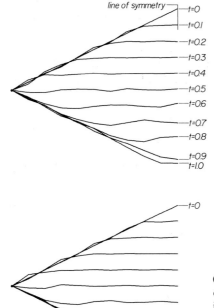

Fig. 10.5 Standing triangular wave in a fixed string of unit length and properties discretized with 10 linear elements for the half of it. Marching in time is stepwise with the explicit scheme (10.32) and $\tau = 0.025$. Compare the computed motion with the theoretical in Fig. 10.1.

Fig. 10.6 Same as Fig. 10.5 but with $\tau = 0.01$. Dispersion is not diminished and we conclude that it is due to the rough discretization in x.

The effect of numerical dispersion on the waveform is clearly visible in Fig. 10.5 (which should be compared with Fig. 10.1) that depicts the computed motion of a fixed (half) string of unit length in which $c = 1$, plucked into a triangular shape and then released. Ten linear two degrees of freedom elements with consistent mass matrices are used to discretize the half string, and the marching in time is done with the explicit scheme (10.32) over time steps $\tau = 0.025$, a step size that is close to the limit of stability. (Since there are 20 elements for the complete string and $h = \frac{1}{20}$, Eq. (10.24) predicts a limit of stability at $\tau = 0.029$.) A smaller time step $\tau = 0.01$ in the same time scheme applied to the same string does not visibly reduce the dispersion as can be seen in Fig. 10.6, indicating that this dispersion, or deformation, of the standing wave is mainly due to the approximation in x, not to that in the time.

Refinement of the mesh causes the lower portion of the eigensystem, the one that usually has the most weight in the approximation of the wave form, to be more precisely computed and as $\hat{\omega}_j \to \omega_j$ in this portion, dispersion recedes. That this is actually the case is seen in Fig. 10.7, showing the same

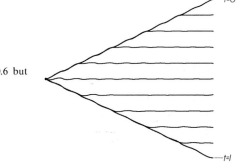

Fig. 10.7 Same as Figs. 10.5 and 10.6 but with 40 linear elements and $\tau = 0.005$.

string as in Figs. 10.5 and 10.6, but modeled now with 40 elements for the half of it, and followed in time with the time step $\tau = 0.005$.

4. Effects of (Numerical) Viscosity

When the string moves through a resisting medium that exerts on it a restraining force proportional to the velocity, its equation of motion becomes

$$\frac{\partial^2 y}{\partial t^2} + q\frac{\partial y}{\partial t} = c^2\frac{\partial^2 y}{\partial x^2} \qquad (10.37)$$

Friction affects the wave propagation in two basic ways: It *attenuates* the wave, robbing mechanical energy from it and thereby reducing its amplitude, and it also disperses the wave. Attenuation is exponentially faster for the high frequency shorter wavelength modes, and hence the visible effect of friction is to smooth the waveform, in opposition to the time and space discretization caused dispersion that exposes the high frequency waves in the wave profile. Viscosity may be artificially introduced through the conditionally stable scheme

$$y_1 = y_0 + \tau\dot{y}_0 + \tfrac{1}{2}\tau^2\ddot{y}_0$$
$$\dot{y}_1 = \dot{y}_0 + \tau[\alpha\ddot{y}_0 + (1 - \alpha)\ddot{y}_1] \qquad (10.38)$$

discussed in Chapter 9, Section 3, with $\alpha < \tfrac{1}{2}$.

The marked smoothing of the waveform resulting from artificial viscosity is seen in Fig. 10.8, which depicts the same string as in Figs. 10.5 and 10.6, discretized with the same 10 linear finite elements for its half, but integrated this time with the explicit scheme (10.38) with $\alpha = 0.35$ and with the time step $\tau = 0.025$.

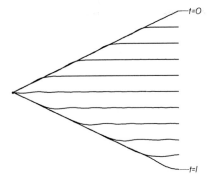

Fig. 10.8 Effect of artificial viscosity introduced by the explicit scheme (10.38) with 10 linear elements ($Ne = 10$), $\alpha < 0.50$ and $\tau = 0.35$. Fast attenuation of the high frequency modes gives the string a smoother appearance.

5. Higher-Order Elements

From the eigenvalue error analysis of Chapter 6, we know that when the string is discretized with polynomial finite elements of order p, then the error in the computed natural frequencies $\hat{\omega}_j$ is $O((h\omega_j)^{2p})$. Hence, with the same number of elements the lower portion of ω_j is more precisely computed with the higher-order elements and the spurious dispersion is reduced.

In the present section we wish, first, to see more precisely how $|\omega_j - \hat{\omega}_j|$ depends on j, and then to numerically observe how the higher-order elements help in reducing the effect of spurious dispersion. To compare $|\omega_j - \hat{\omega}_j|$ for the different finite elements, we compute it for the reference problem of string fixed at both its ends. Let this string be discretized by first-order elements and a *lumped* mass matrix. In this case, $M = I$ and the global (stiffness) matrix is the one in Eq. (6.36). A closed form expression is readily obtained for the jth computed frequency $\hat{\omega}_j$, $\hat{\omega}_j = 2[\sin(\pi jh/2)]/h$, and hence

$$\frac{\hat{\omega}_j}{\omega_j} - 1 = \frac{\sin \xi}{\xi} - 1, \qquad \xi = \pi jh/2, \qquad j = 1, 2, \dots, N \qquad (10.39)$$

Since $h = 1/(N + 1)$ as $N \to \infty$, $\xi \to 0$ for $j = 1$ and $\xi \to \pi/2$ for $j = N$. When a *consistent* mass matrix is used in the finite element discretization, we obtain by the technique of Chapter 6, Section 3, that

$$\frac{\hat{\omega}_j}{\omega_j} - 1 = \frac{\sqrt{3} \sin \xi}{\xi(1 + 2\cos^2 \xi)^{1/2}}, \qquad 0 < \xi < \pi/2 \qquad (10.40)$$

The variation of $(\hat{\omega}/\omega) - 1$ with ξ for the fixed string, discretized with a large number of quadratic and cubic C^0 and C^1 finite elements is shown in Figs. 10.9–10.11. Both the quadratic and cubic C^0 elements have two distinct branches on the $(\hat{\omega}/\omega) - 1$ versus ξ graph, owing to the two distinct shape functions for the internal and interelement nodal points. It is interesting that such discontinuity is barely visible for the C^1 cubic element. Figure 10.12

Fig. 10.9 Discretization error in the computed natural frequencies $\hat{\omega}$ in a fixed string discretized with a very large number of quadratic elements. The two branches belong to the two different basis functions for the interior and the interelement nodes.

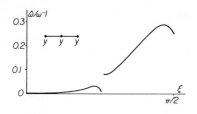

Fig. 10.10 Discretization error in the computed frequencies of a fixed string discretized with cubic C^0 elements.

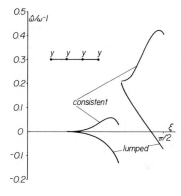

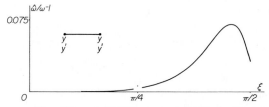

Fig. 10.11 Same as Figs. 10.9 and 10.10 but now the string is discretized with cubic C^1 elements.

Fig. 10.12 Comparison of the frequency errors caused by the elements: (a) linear with lumped mass matrix; (b) linear, consistent: (c) quadratic, consistent; (d) quadratic, lumped; (e) cubic four nodal point C^0, consistent; (f) cubic C^0, lumped; and (g) cubic C^1, consistent. The last reduces spurious dispersion most.

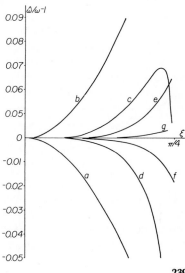

239

compares the error in the natural frequencies computed for the fixed string with linear, quadratic, and cubic elements with and without lumping, and the cubic C^1 element stands out as the most accurate.

To numerically demonstrate the superiority of the cubic C^1 string element in reducing the spurious dispersion we undertake the recomputation of the triangularly plucked string of Figs. 10.5–10.7, using now ten cubic C^1 elements for its half. Figure 10.13 shows the result of this computation which in the time was done with the *implicit* scheme (10.35) with $\tau = 0.005$, and indeed the effect of the spurious dispersion is seen to be substantially reduced. To achieve this reduction of the dispersion we had to use not only high-order elements but also a small time step to minimize the ill effects of the time scheme. A larger time step may reintroduce dispersion as in Fig. 10.14, which shows the same string as that in Fig. 10.13, computed with the same space and time schemes, but with $\tau = 0.05$.

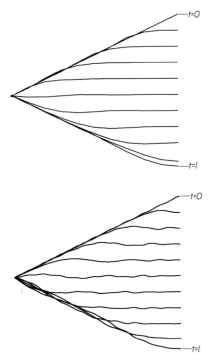

Fig. 10.13 The string of Figs. 10.5–10.7 discretized with 10 cubic C^1 elements and followed in time with the implicit scheme (10.35) with step size $\tau = 0.005$. Dispersion is reduced.

Fig. 10.14 Same as Fig. 10.13 but with the larger $\tau = 0.05$. Dispersion is increased by the time scheme.

Had we chosen the explicit time scheme (10.32), we would have been limited in the choice of τ, according to Eq. (10.26), to $\tau < 0.0075$. But this is still above the time step size $\tau = 0.005$ we used in Fig. 10.3 to reduce the time dispersion below that for the space discretization with cubic C^1 elements, and we could have used the much more convenient explicit scheme instead. In any event, this stepwise integration of the string equation requires at least 200 vector matrix multiplications to go from $t = 0$ to $t = 1$, at which time the string theoretically reassumes the form of an upside-down triangle. Modal decomposition, which we have discussed in Chapter 9, Section 5, appears here to be economically more attractive, and it becomes more so the longer we follow the motion in time since in modal analysis, the algebraic eigenvalue solution of $Kx = \lambda Mx$ is done only once, and the matrices here are relatively small in size.

Traveling waves reveal the effects of spurious dispersion in a more striking manner, as the following numerical examples show. In the first, half of the unit string is discretized with 40 cubic C^1 elements and the initial form

$$y(x, 0) = 1 + \cos(\pi x/4h), \qquad 0 \leqslant x \leqslant 4h$$
$$y(x, 0) = 0, \qquad\qquad\quad 4h \leqslant x \leqslant 0.5 \tag{10.41}$$

where $x = 0$ is at the center of the string, is given to it. After its release, the initial cosine wave separates into two smaller ones moving in opposite directions. Figures 10.15 and 10.16 show the computed motion of the right half of the string between the times $t = 0$ and $t = 1$. In both figures, the string's motion in time is followed stepwise with the *implicit* scheme (10.35), once with $\tau = 0.005$ and then with the smaller $\tau = 0.001$, which considerably reduces the visible dispersion. Here, dispersion is mainly due to the time scheme. We also note with interest that computation with the implicit scheme causes a dispersion that exhibits itself in the form of *lagging* small high frequency waves.

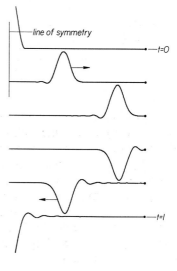

Fig. 10.15   A cosine wave traveling on a string is reflected from the fixed right-hand point. Discretization in x is done with 40 cubic C^1 elements and the wave is followed in time with the implicit scheme (10.35) with time step $\tau = 0.005$. Notice the small lagging waves.

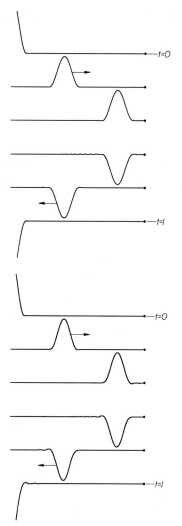

Fig. 10.16 Same as Fig. 10.15 but with the smaller time step $\tau = 0.001$. Disappearance of the visible effects of spurious dispersion indicates that its cause is the time scheme.

Fig. 10.17 Same wave propagation problem as in Fig. 10.16 except that the integration in time is done with the explicit scheme (10.32) with $\tau = 0.0025$. Notice that now the spurious small waves are ahead of the main one.

For comparison, we compute the same problem with the *explicit* scheme (10.32) marching in time with steps $\tau = 0.0025$, which are nearly the largest we can take and still have stability. Dispersion is noticed in Fig. 10.17 but now the small spurious waves run *ahead* of the main one. Again the implicit scheme does not offer us decisive economical advantages over the explicit as a prize for inverting a matrix.

As a further example of spurious dispersion, we compute the traveling wave caused by a *blow* to the string. Here, the initial condition of displacements is zero, but we impart an initial velocity to only the central nodal point in order to simulate a string struck over a very small segment around

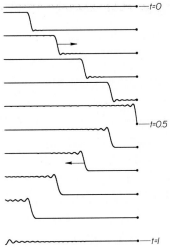

Fig. 10.18 Motion of a string struck over a small area near its center. Discretization in x is done with 40 cubic C^1 elements and the integration in time with the explicit scheme (10.32). The near discontinuity in the displacement is responsible for the strong spurious dispersion.

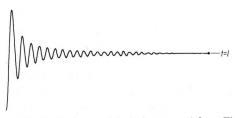

Fig. 10.19 The computed displacement of the string at $t = 1$ from Fig. 10.18 with the displacements greatly enlarged.

its center. Because of the symmetry, only half of the string is considered, and we nodel it with 40 cubic C^1 elements as in the previous examples. In the time, we follow the string with the explicit scheme (10.32) with $\tau = 0.001$. Figure 10.18 shows the traveling wave caused by the central blow and the stresses in the string are seen to be considerable; the wavefront is nearly vertical. Because of the near discontinuity in the string's displacement, the higher modes play a greater role in the approximation and dispersion is more in evidence. Figure 10.19 shows the computed shape of the string, which is theoretically flat, at $t = 1$ with the transverse displacements greatly exaggerated.

6. Spurious Reflection

We have seen in Section 1 how a traveling wave on a string (which could equally well be a surface wave in a water canal) is reflected from a boundary point. Partial reflection occurs also at internal points of discontinuity. If the

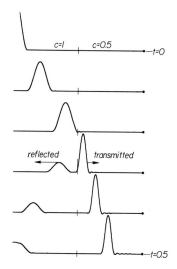

Fig. 10.20 Partial reflection and transmission over a point of discontinuity.

wave equation

$$\frac{\partial^2 y}{\partial t^2} = c^2(x) \frac{\partial^2 y}{\partial x^2} \tag{10.42}$$

has a suddenly changing c due to, say, an abrupt change in the geometrical or material properties of the string or a step in the bed of the water canal, then part of the traveling wave is *transmitted* over it, continuing to travel in its original direction, and another part is *reflected* back. Such a reflection is shown in Fig. 10.20 for the cosine wave of Figs. 10.15–10.17, crossing a point on the string where c changes from $c = 1$ to $c = \frac{1}{2}$.

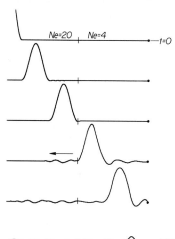

Fig. 10.21 Spurious reflection from a coarse mesh.

A similar spurious reflection occurs when the traveling wave crosses over from a fine mesh region into a coarse mesh region. A given mesh size has a lower limit to the wavelength that it can approximate, or an upper limit ω_N to the frequency it can transmit—the *cutoff* frequency. When the wave enters the coarse mesh region, some of its higher-frequency components cannot penetrate it and are reflected. This can be seen clearly in Fig. 10.21, showing the cosine wave traveling on a string, discretized with 24 cubic C^1 elements, but with the last four elements being five times bigger than the first 20. As the wave reaches the larger elements, small waves appear, traveling backward. Also, because of the larger element size, dispersion becomes more pronounced in the form of leading small waves in front of the original one.

7. Flexural Waves in a Beam

Without damping and with no elastic support, the bending wave equation in the uniform thin elastic beam is

$$\frac{\partial^2 y}{\partial t^2} + c^2 \frac{\partial^4 y}{\partial x^4} = 0 \qquad (10.43)$$

The natural frequencies of the simply supported beam are proportional to $\pi^2 j^2$ not to πj, as is the case for the string, and the beam naturally disperses the flexural waves traveling on it. The eigenfunctions of the unit simply supported beam are known to be $y_j(x) = \sin \pi j x$, with natural frequencies $\omega_j = \pi^2 j^2$. Consequently, corresponding to Eq. (10.17), we have here

$$y(x, t) = \sum_{j=1}^{\infty} \tfrac{1}{2} a_j [\sin \pi j(x - c\pi jt) + \sin \pi j(x + c\pi jt)] \qquad (10.44)$$

and the speed of propagation c_j of the jth eigenmode is

$$c_j = \pi j c \qquad (10.45)$$

Or since $\omega_j = \pi^2 j^2$

$$c_j = c\sqrt{\omega_j} \qquad (10.46)$$

and the higher the mode is, or the smaller its wavelength, the faster its speed. This is physically unacceptable and various more realistic models for the propagation of waves in beams have been suggested that include the shear and rotary inertia that the elementary theory of beams ignores. Taking into account the rotary inertia of the beam results in Rayleigh's equation,

$$c^2 y'''' - \ddot{y}'' + \ddot{y} = 0 \qquad (10.47)$$

which predicts a speed of propagation of the jth mode in the form

$$c_j = c[\omega_j/(\omega_j + 1)]^{1/2}, \qquad \omega_j = \pi^2 j^2 \qquad (10.48)$$

For very long waves (the wavelength λ_j is $\lambda_j = 2\pi/\sqrt{\omega_j}$) compared with the diameter of the beam's cross section $\omega_j \ll 1$ and c_j in Eq. (10.48) approaches that in Eq. (10.46). Otherwise, when $\omega_j \to \infty$, c_j approaches c, not infinity.

If the shear effect is also included in the beam equation, we obtain the *Timoshenko* equation of motion, which has received a great deal of attention in recent years, but we omit this lengthy equation here.

Discretization in x reduces Eq. (10.43) to the system $Ky + M\ddot{y} = 0$ and we can estimate the stable τ, in the explicit scheme (10.32), by solving the element eigenproblem $k_e y_e = \lambda^e m_e y_e$. For the cubic elements,

$$k_e = \frac{1}{h^3}\begin{bmatrix} 12 & 6 & -12 & 6 \\ 6 & 4 & -6 & 2 \\ -12 & -6 & 12 & -6 \\ 6 & 2 & -6 & 4 \end{bmatrix}, \qquad m_e = \frac{h}{420}\begin{bmatrix} 156 & 22 & 54 & -13 \\ 22 & 4 & 13 & -3 \\ 54 & 13 & 156 & -22 \\ -13 & -3 & -22 & 4 \end{bmatrix}$$

$$(10.49)$$

and this element eigenproblem yields $\lambda_4^e = (91.65h^{-2})^2$ here. Consequently, for $Ky = \omega^2 My$, $\omega_N \leqslant 91.65h^{-2}$, and the stable τ is predicted to be limited to

$$0 < \tau < 0.022h^2 \qquad (10.50)$$

While in second-order problems, stability limits τ to only $O(h)$, in the case of the fourth-order beam problem, it limits it to the smaller $O(h^2)$.

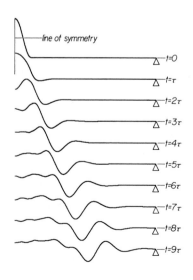

Fig. 10.22 Flexural wave propagation in a beam. Discretization in space is done with 36 cubic elements. The motion of this (half) simply supported unit beam of unit properties is followed in time with the implicit scheme (10.35) with $\tau = 0.0005$.

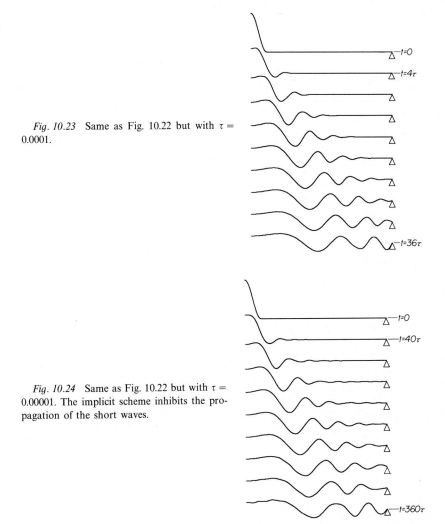

Fig. 10.23 Same as Fig. 10.22 but with $\tau = 0.0001$.

Fig. 10.24 Same as Fig. 10.22 but with $\tau = 0.00001$. The implicit scheme inhibits the propagation of the short waves.

Several numerical examples involving wave propagation on beams are described next. In the first, the simply supported beam is discretized with 36 cubic elements for the half of it, is given the symmetric cosine initial displacement as in Fig. 10.22, and is then released to move as shown in Figs. 10.22–10.24. Its position in time is determined with the *implicit* scheme (10.35) using different step sizes τ. It is interesting that the implicit scheme inhibits the propagation of the higher modes.

Figures 10.25–10.27 show the same beam, but it is now set into motion by a blow at the center.

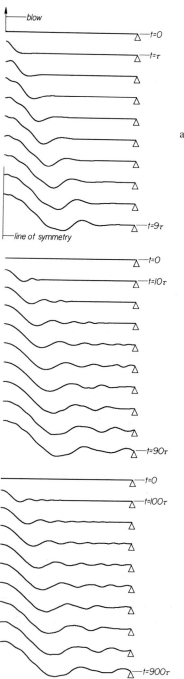

Fig. 10.25   Same beam as in Fig. 10.22 but struck at the center.

Fig. 10.26   Same as in Fig. 10.25 but with a smaller time step $\tau = 0.00005$.

Fig. 10.27   Same as in Fig. 10.26 but with a smaller time step $\tau = 0.000005$.

Notice that Eq. (10.50) predicts a limit $\tau < 4.10^{-6}$ for the above examples, in which $h = \frac{1}{72}$, if computed with the explicit scheme (10.32) and the explicit scheme becomes very expensive. In any event, for the prediction of the beam's motion over long periods of time, modal decomposition is definitely more economical.

8. Stiff String

A real string also has at least some resistance to bending—to curvature change—and when this is taken into consideration, its equation of motion becomes

$$py'' - qy'''' = \rho a \ddot{y}, \qquad 0 < x < l \qquad (10.51)$$

in which p is the tension, q the elastic modulus times the moment of inertia of the cross section, ρ the density, and a the cross section. Piano strings are a good example of stiff strings. The undesirable beam effect is reduced in them by giving a large tension p to the strings in order to render q/p small. (By the way, for musicians the resonance of strings is *sympathetic* vibration.)

Discretization of Eq. (10.51) with finite elements may be carried out with the cubic string and beam elements, resulting as before in $Ky + M\ddot{y} = 0$. In case we wish to solve this with an explicit scheme, the limit of stability on τ in its dependence upon the geometrical, material, and discretization parameters can be obtained conveniently from the element eigenproblem.

EXERCISES

1. Compute the motion of a fixed string when it is elastically supported. The equation of motion here is

   $$y'' = \ddot{y} + y, \qquad 0 < x < l, \qquad y(0) = y(l) = 0$$

   Assume that the string is initially plucked into a triangular form.

2. Repeat Exercise 1 for the nonlinear equation

   $$y''/(1 + y')^2 = \ddot{y}$$

3. Study the effect of (artificial) viscosity on the wave propagation problem shown in Figs. 10.18 and 10.19.

4. Study the effect of (artificial) viscosity on the motion of the centrally struck beam shown in Figs. 10.25–10.27.

5. The equation of motion of a string with rigid pendulums attached along
it is

$$-\frac{\partial^2 \phi}{\partial x^2} + \sin \phi = \frac{\partial^2 \phi}{\partial t^2}, \qquad 0 < x < 1, \quad t > 0$$

where ϕ is the angle through which the pendulums swing. Study the
propagation of a cosine wave in this nonlinear string problem.

6. Solve the *Boussinesq* equation

$$\ddot{y}'' + y'' = \ddot{y}, \qquad 0 < x < 1, \quad t > 0$$

$$y(0) = y(1) = 0$$

for the case of a triangularly plucked string.

7. The fixed stiff string

$$y'' - \varepsilon y'''' = \ddot{y}, \qquad 0 < x < 1, \quad t > 0$$

is struck at the center. Follow the wave motion past reflection and observe the effect of $\varepsilon \to 0$ on it. How does this change when the string is
hit nearer to the fixed end?

8. The element stiffness and mass matrices of a circular cubic C^1 membrane
element between $r_0 \leqslant r \leqslant r_0 + h$ are

$$k_e = \frac{r_0}{30h} \begin{bmatrix} 36 & 3 & -36 & 3 \\ 3 & 4 & -3 & -1 \\ -36 & -3 & 36 & -3 \\ 3 & -1 & -3 & 4 \end{bmatrix} + \frac{1}{60} \begin{bmatrix} 36 & 6 & -36 & 0 \\ 6 & 2 & -6 & -1 \\ -36 & -6 & 36 & 0 \\ 0 & -1 & 0 & 6 \end{bmatrix}$$

and

$$m_e = \frac{hr_0}{420} \begin{bmatrix} 156 & 22 & 54 & -13 \\ 22 & 4 & 13 & -3 \\ 54 & 13 & 156 & -22 \\ -13 & -3 & -22 & 4 \end{bmatrix} + \frac{h^2}{8400} \begin{bmatrix} 72 & 14 & 54 & -12 \\ 14 & 14 & 14 & -3 \\ 54 & 14 & 240 & -30 \\ -12 & -3 & -30 & 5 \end{bmatrix}$$

Use these to follow the motion of the fixed circular membrane struck over
a small area around its center.

9. Wave propagation in an infinite medium is computationally solved by
confining the discretization to a finite portion of the medium with appropriate boundary conditions that *avoid reflection*. Show by substitution
that

$$y' = (1/\alpha)\dot{y}, \qquad \alpha = \pm c$$

satisfy the wave equation

$$y'' = (1/c^2)\ddot{y}$$

and hence the boundary condition $\alpha y' = \dot{y}$ conforms to the wave slope and will transmit it through the boundary point without reflection. Numerically solve an infinite string problem reducing it to finite size to observe the influence of the spurious dispersion.

Discuss the difficulty in the more general case

$$(p(x)y')' = \ddot{y} \qquad \text{or} \qquad y'' = \ddot{y} + y$$

SUGGESTED FURTHER READING

Morse, P. M., *Vibration and Sound*. McGraw-Hill, New York, 1948.
Lamb, H., *The Dynamical Theory of Sound*. Dover, New York, 1960.
Kolsky, H., *Stress Waves*. Dover, New York, 1963.
Bland, D. R., *Vibrating Strings*. Routledge & Kegan Paul, London, 1965.
Whitham G. B., *Linear and Nonlinear Waves*. Wiley, New York, 1974.

Index

Computer Science and Applied Mathematics

A SERIES OF MONOGRAPHS AND TEXTBOOKS

Editor
Werner Rheinboldt
University of Maryland